T0140580

# Lecture Notes in Electrical Engineering

## Volume 379

*About this Series*

"Lecture Notes in Electrical Engineering (LNEE)" is a book series which reports the latest research and developments in Electrical Engineering, namely:

- Communication, Networks, and Information Theory
- Computer Engineering
- Signal, Image, Speech and Information Processing
- Circuits and Systems
- Bioengineering

LNEE publishes authored monographs and contributed volumes which present cutting edge research information as well as new perspectives on classical fields, while maintaining Springer's high standards of academic excellence. Also considered for publication are lecture materials, proceedings, and other related materials of exceptionally high quality and interest. The subject matter should be original and timely, reporting the latest research and developments in all areas of electrical engineering.

The audience for the books in LNEE consists of advanced level students, researchers, and industry professionals working at the forefront of their fields. Much like Springer's other Lecture Notes series, LNEE will be distributed through Springer's print and electronic publishing channels.

More information about this series at http://www.springer.com/series/7818

Ping Jack Soh
Hamzah Asyrani Sulaiman
Mohd Azlishah Othman
Mohamad Zoinol Abidin Abd. Aziz
Mohd Fareq Abd Malek
Editors

# Theory and Applications of Applied Electromagnetics

## APPEIC 2015

 Springer

*Editors*
Ping Jack Soh
University Malaysia Perlis
Arau
Malaysia

Hamzah Asyrani Sulaiman
Universiti Teknikal Malaysia Melaka
Durian Tunggal
Malaysia

Mohd Azlishah Othman
Universiti Teknikal Malaysia Melaka
Durian Tunggal
Malaysia

Mohamad Zoinol Abidin Abd. Aziz
Universiti Teknikal Malaysia Melaka
Hang Tuah Jaya
Durian Tunggal
Malaysia

Mohd Fareq Abd Malek
Kampus Kubang Gajah
Universiti Malaysia Perlis
Arau
Malaysia

ISSN 1876-1100          ISSN 1876-1119  (electronic)
Lecture Notes in Electrical Engineering
ISBN 978-3-319-80727-0       ISBN 978-3-319-30117-4   (eBook)
DOI 10.1007/978-3-319-30117-4

This Springer imprint is published by Springer Nature
The registered company is Springer International Publishing AG Switzerland

# Preface

The second Applied Electromagnetic International Conference (APPEIC 2015) was held in the Krabi southern province of Thailand between 3 and 5 December 2015. APPEIC 2015 was organized by the Malaysia Technical Scientist Association (MALTESAS) in collaboration with Universiti Malaysia Perlis (UNiMAP). The conference focused on current developments related to applied electromagnetics, such as antenna, microwave components, and wave propagation. APPEIC 2015 was able to attract participants from Malaysia, Indonesia, Thailand, Hong Kong, Korea, India, United Kingdom, Philippines, and Egypt.

Two outstanding keynote speeches were presented at APPEIC 2015. The first was delivered by Dr. Mohd Haizal Bin Jamaluddin, Senior Researcher from the well-known Wireless Communication Center (WCC) in Universiti Teknologi Malaysia (UTM). Dr. Haizal presented the latest developments in dielectric resonator antennas (DRAs) for future wireless communication (5G). The second keynote speech was delivered by Associate Prof. Dr. Badrulhisham Ahmad, Chairman of IEEE AP/MTT/EMC Malaysian chapter, who reviewed recent developments concerning the popular topic of metamaterial structures, focusing on split ring resonators (SRR).

This book is a compilation of interesting articles on scientific theories and applications in the field of applied electromagnetics. All selected articles were peer reviewed by experts in applied electromagnetics. The editors wish to express their gratitude to MALTESAS and UNiMAP and to Narujaya Consultant for organizing and supporting this very successful conference.

Ping Jack Soh
Hamzah Asyrani Sulaiman
Mohd Azlishah Othman
Mohamad Zoinol Abidin Abd. Aziz
Mohd Fareq Abd Malek

# Contents

# Analytical Modeling for Switched Damping Electromagnetic Energy Harvester

**Beng Lee Ooi, James M. Gilbert, A. Rashid A. Aziz and Chung Ket Thein**

**Abstract** This paper presents a novel switched damping method which is able to improve the operational bandwidth of a linear electromagnetic energy harvester. In this system, different damping coefficients, in which values are computed based on the adaptive percentage-varying factor, are electronically switched at different points within the predefined oscillating quadrants. An analytical model of the switching damping device is presented, and the predictive results indicate that a reasonable large range of frequency shifts can be achieved for the proposed output resistances. The outputs for a range of values of internal coil resistance and electromotive force (emf) constant are recorded, and for the proposed system, the frequency can be shifted up to range of $+19.50$ to $-81.46$ % away from the initial resonance.

**Keywords** Vibration energy harvesting · Electromagnetic induction · Active switching device · Switching damping system · Frequency tuning

B.L. Ooi (✉)
Faculty of Integrative Sciences and Technology (FIST),
Quest International University Perak, 227 Jalan Raja Permaisuri Bainun,
30250 Ipoh, Perak, Malaysia
e-mail: benglee_85@hotmail.com

J.M. Gilbert
School of Engineering University of Hull, Cottingham Road, Hull HU6 7RX, UK

A.R.A. Aziz
Center for Automotive Research and Electric Mobility (CAREM),
Universiti Teknologi PETRONAS (UTP), Bandar Seri Iskandar,
31750 Tronoh, Perak, Malaysia

C.K. Thein
School of Engineering, Taylor's University, Taylor's Lakeside Campus,
no. 1, Jalan Taylor's, 47500 Subang Jaya, Selangor, Malaysia

© Springer International Publishing Switzerland 2016
P.J. Soh et al. (eds.), *Theory and Applications of Applied Electromagnetics*,
Lecture Notes in Electrical Engineering 379, DOI 10.1007/978-3-319-30117-4_1

1

# 1 Introduction

Energy harvesting is the process to acquire energy derived from environmental sources (e.g., solar, heat, wind, and vibration), captured and stored for use in small electronic devices, such as wearable electronics and wireless sensor networks. Enable the energy harvesting capability to the devices by unlocking the potential for wireless networks to be self-sustainable and virtually perpetual operation which is limited by the hardware parts rather than the finite energy sources, such as batteries [1]. Due to the availability of vibration energy in most of the targeted wireless sensor node location, it has been suggested as one of the most highly potential energy sources to be used for self-powered portable devices [2]. Over the past few years, many linear vibration-to-electricity converter devices have been investigated for sourcing energy to the wireless devices perpetually [3, 4]. However, regardless of the types of transducer used piezoelectric [5], magnetostrictive [6], electrostatic [7], or electromagnetic [8], linear conversions will only provide the optimum output during the frequency-matched condition. When the external driving frequency is deviated away from the resonance, the generated output will be dropped significantly and it became relatively low compared to the output generated in the resonant mode, which is too little to be utilized practically. In most of the applications, the source frequencies at the targeted location for the devices may not be able to quantify since it constantly varies over time. Therefore, it is clearly beneficial to implement an energy converter that is designed to operate over a wider range of source frequencies.

Several researchers have actively seeking for alternative solutions to address the frequency-matching problem, which includes frequency tuning [9], magnetorheological elastomer [10], frequency up-conversion (FupC) [11], cantilever array [12], nonlinear oscillation [13], and dual resonator [14]. All the proposed methods have a fixed range of operating frequencies after configuration, and none of the previous literature works demonstrate how to shift the dominant frequencies of the harvester devices by actively changing the effective electrical damping of the system while the resonator is still oscillating. This paper proposes a novel method to tune the resonant frequency above or below the initial resonance by applying different resistive load values to the system at various phases during the oscillating cycle. The background and model of a vibration-based conversion are presented, followed by the analytical model of a switched damping system. The values of the load resistance are computed based on the percentage-varying adaptation method. The predictive shifted resonant frequencies for a range of values of coil resistances and emf constants are plotted and compared with the resonant frequency of a fixed damping system. Lastly, the results are discussed and conclusions drawn.

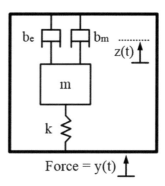

## 2 Background of Vibration-Based Conversion

Vibration-based energy harvesters can be generally modeled as a mass, spring, and damper system of the type illustrated in Fig. 1 [15], where $m$ is the seismic mass, $k$ is the spring constant, and $b_e$ and $b_m$ are the electrically induced and mechanical damping coefficients, respectively. The energy that is dissipated in the $b_m$ damper represents the mechanical losses, whereas the energy that is dissipated in the $b_e$ damper represents the available energy from vibration force that is successfully converted into electrical domain. The equation of motion of the system can be derived as follows:

$$m\ddot{z}(t) + (b_m + b_e)\dot{z}(t) + kz(t) = -m\ddot{y}(t) \tag{1}$$

where $z(t)$ is the time-varying position of the seismic mass relative to the casing, which is driven by the time-varying sinusoidal force, $y(t)$. Assuming a circuit with an internal coil resistance, $R_i$, and an output resistive load resistance, $R_L$, then the electrically induced damping coefficient can be determined as follows [16]:

$$b_e = \frac{K_e^2}{R_i + R_L} \tag{2}$$

where $K_e = NBl_e$ is the emf constant, $N$ is the number of turns for a pickup coil, and $l_e$ is the effective length of the coil that is perpendicular to the direction of the magnetic field $B$.

Given that the vibrating source is described by $y(t) = Y\sin(\omega t)$, where $\omega$ is the external driving frequency and $Y$ is the amplitude of the source displacement, the available average electrical power applied into electrical domain can be given as follows [17]:

$$P_{avg} = \frac{m\zeta_e A^2(\omega^2/\omega_n^3)}{[1 - (\omega/\omega_n)^2]^2 + [2\zeta(\omega/\omega_n)]^2} \tag{3}$$

where $\omega_n = \sqrt{k/m}$ is the resonant frequency of the system, $A$ is the amplitude of the source acceleration which can be derived from $Y = A/\omega^2$, and $\zeta = \zeta_m + \zeta_e = (b_m + b_e)/2m\omega_n$ is the damping factor of the system, which is also the sum of mechanical ($\zeta_m$) and electrically induced ($\zeta_e$) damping factors. The electrically induced damping factor may be then given as follows:

$$\zeta_e = \frac{K_e^2}{2m\omega_n(R_i + R_L)}.$$ (4)

Given that all the mechanical and parasitic losses in the system are fixed after the configuration of the system, this clearly indicates that changes in electrically induced damping factor can be achieved by altering the value of the resistive load resistance, $R_L$, similar to the damping factor of the system, $\zeta$.

It may also noteworthy to mention that with a higher internal coil resistance, $R_i$, this will create a greater loss in the circuit, hence compromising the output of the system. However, due to the practical coil limitation, it is not possible to fully neglect the $R_i$, so to achieve a greater performance, one should always design an energy harvesting system based on the lowest possible value of $R_i$ and the greatest possible value of emf constant, $K_e$, which will be covered in Sect. 4.

## 3   Modeling of Switched Damping System

Aforementioned analysis gives the model of an energy harvesting device with a fixed damping that is subjected to an external vibration. However, in this section, a model of switched damping system is presented and the effect toward the resonant frequency is analyzed. In this system, a different effective electrically induced damping is applied at different stages during the oscillation cycle. Assume that the cycle is divided into four quadrants which are defined according to the displacement, $z$ and velocity, $\dot{z}$ of the resonator, as illustrated in Fig. 2. Quadrant I is when the resonator beam moves to the equilibrium from the maximum position, and quadrant II is when the beam moves further downward to the minimum position from the equilibrium. Quadrant III is when the beam changes the direction and

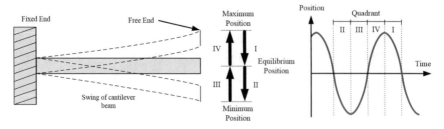

**Fig. 2** Motion and direction of the electromagnetic cantilever beam per quadrants

moves upward to the equilibrium from the minimum position, and lastly, quadrant IV is when the beam moves further upward to the maximum position from the equilibrium. To avoid complex analysis, these quadrants may be applied with the electrically induced damping according to:

$$b_e = \begin{cases} b_{e1} & \text{when } z \cdot \dot{z} < 0 \\ b_{e2} & \text{when } z \cdot \dot{z} \geq 0. \end{cases} \tag{5}$$

Damping coefficient $b_{e1}$ is applied in quadrants I and III, while damping coefficient $b_{e2}$ is applied in quadrants II and IV. To further illustrate the effects of different damping regimes in different quadrants, the velocity $\dot{z}$ against displacement $z$ phase are plotted in Fig. 3 to show the differences between higher damping, lower damping and switched damping for case when $b_{e1} < b_{e2}$.

To change the resonant frequency of an electromagnetic energy harvester, one can alternately switch the effective output load of the device at different oscillation quadrants. This can be proved analytically by computing the duration required by the system to complete each quadrant in Fig. 3c. It can be noticed that the duration

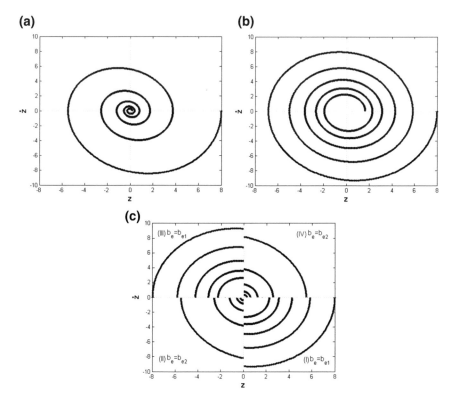

**Fig. 3** Phase plots for damping system. **a** Higher damping, **b** lower damping, and **c** switched damping ($b_{e1} < b_{e2}$)

required is similar in quadrants I and II and in quadrants II and IV, so only two types of quadrant need to be considered in this model and they are quadrant I when $z > 0$ and $\dot{z} < 0$ and quadrant II when $z < 0$ and $\dot{z} < 0$. With the absence of external driving input, the time when the trajectory leaves quadrant I (with condition $z(t) = 0$) and quadrant II (with condition $\dot{z}(t) = 0$) has been determined as (6) and (7), respectively [18]:

$$t_{\mathrm{I}} = t|_{z(t)=0} = \frac{\pi - \tan^{-1}\left[\sqrt{(1 - \zeta_1^2)}/\zeta_1\right]}{\omega_n\sqrt{1 - \zeta_1^2}} \tag{6}$$

$$t_{\mathrm{II}} = t|_{\dot{z}(t)=0} = \frac{\tan^{-1}\left[\sqrt{(1 - \zeta_2^2)}/\zeta_2\right]}{\omega_n\sqrt{1 - \zeta_2^2}} \tag{7}$$

where $\zeta_1 = (b_m + b_{e1})/2\omega_n m$ and $\zeta_2 = (b_m + b_{e2})/2\omega_n m$ are the damping factors for quadrants I and III and quadrants II and IV, respectively.

The values for $\zeta_1$ and $\zeta_2$ may be determined in number of ways. In this study, the damping factors are selected according to the resistive load that is attached to the system as follows:

$$\zeta_1 = \zeta m + \zeta_{e1} = \zeta_m + \frac{K_e^2}{2m\omega_n(R_i + R_{L1})} \tag{8}$$

$$\zeta_2 = \zeta m + \zeta_{e2} = \zeta_m + \frac{K_e^2}{2m\omega_n(R_i + R_{L2})} \tag{9}$$

where $R_{L1}$ and $R_{L2}$ are the switching load resistances that are attached to the system during quadrants I and III and quadrants II and IV, respectively. As in the low-frequency system, the effects of the inductance in the generating circuit are negligible as the resistance effect is always dominating, and hence, to avoid any complex conjugate in these analyses during the impedance-matching condition, the load-matching-condition resistance is assumed to be equal to the internal coil resistance ($R_m = R_i$) unless otherwise stated. The load resistance $R_{L1}$ and $R_{L2}$ may be then modified by scaling their values up or down from the load-matching-condition resistor, $R_m$, adaptively by a varying factor, $v_F$, according to one of the three cases:

Case I is when $R_{L1} > R_{L2}$, and this can be achieved by adding the resistance $R_m v_F$ to $R_{L1}$ and deducting it from $R_{L2}$ as follows:

$$\text{Selected resistances for Case I} = \begin{cases} R_{L1} = R_m + (R_m F) \\ R_{L2} = R_m - (R_m \times v_F) \end{cases} \tag{10}$$

Case II is when $R_{L1} < R_{L2}$, and this can be achieved by deducting the resistance $R_m v_F$ to $R_{L1}$ and adding it to $R_{L2}$ as follows:

$$\text{Selected resistances for Case II} = \begin{cases} R_{L1} = R_m - (R_m \times v_F) \\ R_{L2} = R_m + (R_m \times v_F) \end{cases} \quad (11)$$

and lastly, Case III is when $v_F = 0\%$ and $R_{L1} = R_{L2} = R_m$ which leads to $\zeta_1 = \zeta_2 = \zeta$, the fixed damping system condition.

According to the resistance adaptation given in (10) and (11), it is possible to determine the complete cycle duration, $T$, for values of $v_F$ equal to 0, 5, 10, 90, and 95 % analytically from (6) to (7). Therefore, the shifted resonant frequency that is caused by alternating the resistive loads in the switching device can be determined by $\omega_d = 2\pi/T$, where the period, $T = 2(t_1 + t_{II})$, for one complete cycle of the oscillation by the system may be determined as follows:

$$T = \frac{2}{\omega_n} \left[ \frac{\pi - \tan^{-1}\left[\sqrt{(1-\zeta_1^2)}/\zeta_1\right]}{\sqrt{1-\zeta_1^2}} + \frac{\tan^{-1}\left[\sqrt{(1-\zeta_2^2)}/\zeta_2\right]}{\sqrt{1-\zeta_2^2}} \right]. \quad (12)$$

For a system with a fixed damping ($\zeta_1 = \zeta_2 = \zeta$), it is obvious that the period from (12) can be derived into $T = 2\pi/(\omega_n\sqrt{1-\zeta^2})$, which is the duration for a resonator in a fixed damping system to complete the full cycle.

## 4   Results and Discussion

The normalized shifted resonant frequencies as a function of $R_{L1}$ and $R_{L2}$ are plotted for a range of values of internal load resistances $R_i$ and emf constant $K_e$ as shown in Fig. 4, where $\omega_d$ is the shifted resonant frequency and $\omega_n$ is the initial system resonant frequency. With an arbitrarily selected mechanical damping as $\zeta_m = 0.04$, the normalized frequency, $\omega_d/\omega_n$, is computed based on the values of parameter as given in Table 1. All the values are selected based on the assumption made to emulate a practical electromagnetic energy harvesting device with a large number of coil turns and emf constant.

From Fig. 4, it shows three shifted resonant frequency plots for a range of values of internal load resistances when $R_i = 20$, 30, and 40 $\Omega$, respectively. Three different values of emf constants were also investigated, and they are $K_e = 5$, 8, and 10. It can be seen that the resonant frequency can be adjusted above or below the resonance of the fixed damping system and a reasonably large range of frequencies can be achieved. The achievable range of shifted resonant frequencies for range of values of $R_i$ and $K_e$ are simulated and summarized in Table 2. From the results, up to +19.50 % for $\omega_d > \omega_n$ (upward) and −81.46 % for $\omega_d < \omega_n$ (downward), frequency shifting is recorded in the given test case when $R_i = 20 \Omega$ and $K_e = 10$,

**Fig. 4** Effect of relative load resistances $R_{L1}$ and $R_{L2}$ on switched damping system resonant frequency for **a** $R_i = 20\ \Omega$, **b** $R_i = 30\ \Omega$, and **c** $R_i = 40\ \Omega$

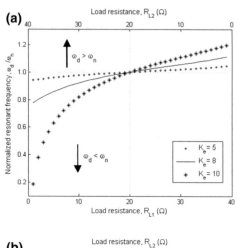

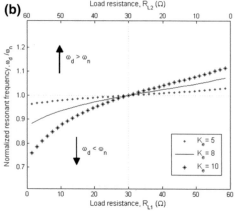

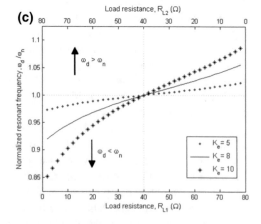

**Table 1** Simulation parameters of switched damping device

| Parameters | Values for range of internal coil resistances | | | Units |
|---|---|---|---|---|
| | $R_i = 20\,\Omega$ (Fig. 4a) | $R_i = 30\,\Omega$ (Fig. 4b) | $R_i = 40\,\Omega$ (Fig. 4c) | |
| $R_m$ | 20 | 30 | 40 | $\Omega$ |
| $\zeta_m$ | 0.04 | | | – |
| $\omega_n$ | 125.7 | | | rad/s |
| $m$ | 20 | | | gram |
| $K_e$ | 5 & 8 & 10 | | | Vs/m |
| $R_{L1}$ | 1–39 | 1.5–58.5 | 2–78 | $\Omega$ |
| $R_{L2}$ | 39–1 | 58.5–1.5 | 78–2 | $\Omega$ |
| $v_F$ | 0–90 (step of 5) | | | % |

**Table 2** Summary of achievable range of shifted resonant frequencies on switching damping system for $R_i = 20$, 30, and 40 $\Omega$

| Internal load resistance, $R_i(\Omega)$ | Emf constant, $K_e$ | | | | | |
|---|---|---|---|---|---|---|
| | $K_e = 5$ | | $K_e = 8$ | | $K_e = 10$ | |
| | $\omega_d > \omega_n$ (%) | $\omega_d < \omega_n$ (%) | $\omega_d > \omega_n$ (%) | $\omega_d < \omega_n$ (%) | $\omega_d > \omega_n$ (%) | $\omega_d < \omega_n$ (%) |
| 20 | +4.40 | −5.89 | +11.10 | −22.49 | +19.50 | −81.46 |
| 30 | +3.00 | −3.69 | +7.30 | −11.71 | +11.60 | −24.21 |
| 40 | +2.30 | −2.70 | +5.50 | −7.97 | +8.50 | −14.79 |

respectively. It may be noted that the values of $K_e$ directly correspond to the achievable range of shifted resonant frequencies, whereas the values of $R_i$ inversely correspond to it. This is achieved by decreasing the $R_i$ and increasing the $K_e$, and this may result in a greater damping factor attached to the system as suggested by (8) and (9), which affect the duration for each oscillating quadrant ($t_I$ and $t_{II}$). Hence, a more significant range of shifted resonant frequencies can be recorded. It may also noteworthy to mention that smaller $R_i$ will lead to lower electrical losses in the circuit, and therefore, greater output power can be delivered to the electrical load.

Besides, it can be noticed that the frequency shifting is greater in $\omega_d < \omega_n$ region as compared to the shifting in $\omega_d > \omega_n$ region. This is caused by the nonlinearity behavior of the values of damping factors $\zeta_1$ and $\zeta_2$ that are selected according to the load resistances $R_{L1}$ and $R_{L2}$ as derived in (8) and (9), which are computed by a varying factor $v_F$. Moreover, at lower $\omega_d$, this causes the cantilever beam to oscillate at higher displacement amplitude as compared to higher $\omega_d$, and hence, changes in the beam's velocity at lower $\omega_d$ cause a greater frequency shifting in $\omega_d < \omega_n$ region.

To explain in more detail on how the resonant frequency has been shifted, Fig. 5 illustrates the dynamic motion plots of switched damping systems for which (a) when $b_{e1} < b_{e2}$ and (b) when $b_{e1} > b_{e2}$, respectively. Say that both of the

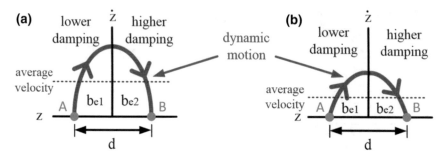

**Fig. 5** Phase plots of dynamic motion in the switched damping system, **a** Case I when $b_{e1} < b_{e2}$ and **b** Case II when $b_{e1} > b_{e2}$

oscillating resonators are required to move from position A to B. For a system with $b_{e1} < b_{e2}$, the resonator goes into the lower damping quadrant and the higher damping quadrant later. Reversely for a system with $b_{e1} > b_{e2}$, the resonator will start with the higher damping quadrant and the lower damping quadrant later. Although the traveling distance, $d$, covered by both resonators is equal, the average velocity, $\dot{z}$, of the resonator is higher in the system with $b_{e1} < b_{e2}$ than the system with $b_{e1} > b_{e2}$. This implies that for the system in (a), it takes lesser time for the resonator to move from position A to B, and hence, the shifted resonant frequency is greater than the initial resonance or $\omega_d > \omega_n$. On the contrary for the system in (b), due to the lower average velocity of the resonator, it takes longer time for the resonator to travel for the same distance (position A to B), and hence, the shifted resonant frequency is lesser than the initial resonance or $\omega_d < \omega_n$. Lastly, one should be noted that altering the damping factor affects not only the resonant frequency of the system but also the available output power. However, not to further extend the context, quantifying the output power is set beyond the scope of the paper.

## 5 Conclusion

This paper presents a novel method to shift the resonant frequency of an electromagnetic energy harvester by altering the system effective load. The analytical results demonstrate that the proposed method is able to shift the resonant frequency above or below the initial resonance by using the resistance adaptation method given in (10), (11), and (12). For a system with an initial resonant frequency of 125.7 rad/s, the output results computed from the model have successfully demonstrated that the switched damping device is able to shift the resonant frequency up to +19.50 % for $\omega_d > \omega_n$ (upward) and −81.46 % for $\omega_d < \omega_n$ (downward) regions, thus widening the operational bandwidth of the system. The results also illustrate that by minimizing the internal coil resistance $R_i$ and maximizing the emf constant $K_e$, greater resonant frequency deviation from the initial resonance can be achieved. Lastly, it is also noteworthy to mention that frequency tuning of this switched damping device can be achieved in real time by altering the load resistance during oscillation without the need to suspend the harvesting process. However, the drawback is that, other than the resonant frequency, the available output power will also be affected. Unfortunately, at this stage, there is no close expression to model the power generated from the device. Hence, the future work will mainly focus on the derivation of a model for average power generated from the system.

**Acknowledgments** The authors would like to thank to all the collaborators and involved parties for the facilities, equipment, and technical support in this project, which were also funded by the MOE-FRGS Grant No. FRGS/1/2014/TK03/QUEST/03/1.

# References

1. Ulukus S, Yener A, Erkip E, Simeone O, Zorzi M, Grover P, Huang K (2015) Energy harvesting wireless communications: a review of recent advances. IEEE J Sel Areas Commun 33(3):360–381
2. Ooi BL, Thein CK, Yew CK, Aziz ARA (2015) Book chapter: applications of vibration-based energy harvesting (veh) devices. IGI Global (In press)
3. Galchev T, Hanseup K, Najafi K (2011) Micro power generator for harvesting low-frequency and nonperiodic vibrations. J Microelectromech Syst 20(4):852–866
4. Ooi BL, Aziz ARA, Gilbert JM (2014) Analytical and numerical modelling for dual-resonator vibration-based electromagnetic generator. In: IEEE 2014 5th international conference on intelligent and advanced systems (ICIAS), pp 1–5
5. Kim GW (2015) Piezoelectric energy harvesting from torsional vibration in internal combustion engines. Int J Automot Technol 16(4):645–651
6. Wang L, Yuan FG (2008) Vibration energy harvesting by magnetostrictive material. Smart Mater Struct 17(4):45009
7. Basset P, Galayko D, Cottone F, Guillemet R, Blokhina E, Marty F, Bourouina T (2014) Electrostatic vibration energy harvester with combined effect of electrical nonlinearities and mechanical impact. J Micromech Microeng 24(3):035001
8. Beeby SP, Torah RN, Tudor MJ, Glynne-Jones P, O'Donnell T, Saha CR, Roy S (2007) A micro electromagnetic generator for vibration energy harvesting. J Micromech Microeng 17(7):1257
9. Aboulfotoh NA, Arafa MH, Megahed SM (2013) A self-tuning resonator for vibration energy harvesting. Sens Actuators, A 201:328–334
10. Sun W, Jung J, Wang XY, Kim P, Seok J, Jang DY (2015) Design, simulation, and optimization of a frequency-tunable vibration energy harvester that uses a magnetorheological elastomer. Adv Mech Eng 7(1)
11. Zorlu O, Topal ET, Kulah H (2011) A vibration-based electromagnetic energy harvester using mechanical frequency up-conversion method. IEEE Sens J 11(2):481–488
12. Lin SC, Lee BS, Wu WJ, Lee CK (2009) Multi-cantilever piezoelectric mems generator in energy harvesting. Rome (IEEE), pp 755–758
13. Cottone F, Vocca H, Gammaitoni L (2009) Nonlinear energy harvesting. Phys Rev Lett 102:80601
14. Ooi BL, Gilbert JM (2014) Design of wideband vibration-based electromagnetic generator by means of dual-resonator. Sens Actuators, A 213:9–18

15. Roundy S (2005) On the effectiveness of vibration-based energy harvesting. J Intell Mater Syst Struct 16(10):809–823
16. Beeby SP (2006) Mj Tudor, and Nm White. Energy harvesting vibration sources for microsystems applications. Meas Sci Technol 17(12):R175
17. Roundy S (2003) Energy scavenging for wireless sensor nodes with a focus on vibration to electricity conversion. Ph.D. thesis
18. Ooi BL, Gilbert JM, Aziz ARA (2015) Switching damping for a frequency-tunable electromagnetic energy harvester. Sens Actuators, A 234:311–320

# Scalable Remote Water Monitoring System Using Radio Frequency Links

**Gerino Mappatao, Christian Aragones, Paul David Elpa, Dea Marielle Pangan and Queenie Santos**

**Abstract** This paper presents a system that is ideal for water monitoring over stretched distances. The system is composed of a computer-based control monitoring unit (CMU) and several remote process transceivers (RPTs). The CMU contains a software program and a transceiver. It issues commands to and receives sensor data from the RPTs. It also saves previous data and displays the current sensor data from each RPT through a graphical user interface. The RPTs are stationed along the length of water systems like rivers where water data are sampled and sent to the CMU. Sensor data include the temperature, level, and turbidity of the water at the location of a particular RPT. The scalability feature allows the system to cover a much extended distance along the length of a river. It is achieved through the ability of the RPTs to relay the command from the CMU to other RPTs and the data from other RPTs to the CMU using user-defined protocols. Test results show that an effective distance of 1300 m is achieved between the CMU and the nearest RPT and between RPTs. Further, there are no significant differences between the actual sensor data and the values received at the CMU.

**Keywords** Scalable · Data transmission · Water monitoring · Radio frequency link · User-defined protocol

## 1 Introduction

The constant degradation of water systems, like rivers in most parts of the world, calls for close monitoring of water quality in these systems. In the Philippines, water from the rivers is the primary source of domestic water, especially for major urban areas. However, the neglect of existing government policies led to the present

G. Mappatao (✉) · C. Aragones · P.D. Elpa · D.M. Pangan · Q. Santos
De La Salle University, 2401 Taft Avenue, 922 Manila, Philippines
e-mail: gerino.mappatao@dlsu.edu.ph

© Springer International Publishing Switzerland 2016
P.J. Soh et al. (eds.), *Theory and Applications of Applied Electromagnetics*,
Lecture Notes in Electrical Engineering 379, DOI 10.1007/978-3-319-30117-4_2

state where only one-third of the river systems are considered suitable for public water supply [1] due to pollution. The main source of pollution is untreated domestic and industrial wastewater [2], p. 4. According to the studies, the effects of pollution of water systems will continue to be felt in the country in the future. It is estimated that in 2025, water availability in the Philippines will be marginal in most of the major cities and in 8 of the 19 major river basins [2], p. 8. Water pollution also leads to problems in the fishing and tourism industries [3]. Water monitoring systems, especially for river systems, must be capable of covering lengthy and large target areas.

The proposed system monitors water quality remotely using radio frequency links to transmit and receive water data using sensors over long and wide target areas. The main feature of the proposed water monitoring system is its scalability, where more RPTs can be added to increase the coverage area. Further, it makes use of a communication system for the control and reception/transmission of sensor data, instead of using a public switched network as proposed in [4, 5]. All sensors used in this system are developed and calibrated using low-cost and locally available components. Figure 1 shows the general system setup of the proposed water monitoring system.

The monitoring system is composed of a CMU and several RPTs. The number of RPTs depends on the desired coverage area. In its ideal setup, the CMU communicates with RPT1, RPT1 communicates with RPT2, and so on. The CMU sends commands periodically to all RPTs for the latter to send their sensor data to the CMU. Each RPT can act as a relay station for the succeeding RPTs to receive the same command from the CMU. For the setup shown in Fig. 1, the command from the CMU will be received by RPT1 and the same command will be relayed by RPT1 to RPT2 and RPTN will receive the same command through RPT2. In the upstream operation, when RPT1 receives the command from the CMU, it transmits its sensor data directly to the CMU, where the data are stored and displayed. The other RPTs will also follow suit upon reception of the relayed command. However, for RPT2, its sensor data will be sent to and relayed by RPT1 to the MCU. The sensor data of RPTN will be sent to RPT2 and then to RPT1 and eventually to the CMU. If, for any instance, an RPT other than RPT1 receives the command directly from the CMU, the command will be discarded or ignored by that RPT. Likewise, if an RPT nearer to the CMU transmits its sensor data, all RPTs farther than the transmitting RPT will ignore the transmission.

**Fig. 1** System configuration of the monitoring system

## 2   System Design

The proposed water monitoring system using radio frequency links is designed to gather water sensor data from remote stations to a central monitoring station. A transceiver is used both at the CMU and at all RPTs. All RPTs have a micro-controller serving as the host to interface the sensors with the transceiver. Figure 2 illustrates the internal parts of the CMU and RPTs.

## 2.1   Control Monitoring Unit (CMU)

The CMU consists of a transceiver that is connected to the computer via an interface cable. This is where all commands will be emanating for RPTs to send their sensor data. Also, all sensor data from RPTs will be received and displayed at the CMU through a graphical user interface (GUI). These data will be further processed and stored in a database and also at the CMU. At the CMU, a command is automatically and periodically issued by the system at a predetermined time set by the user. Commands can also be sent from the CMU manually by the user though the GUI. The GUI used in this project was developed using Visual Basic .Net 2005. The CMU is basically composed of a computer and a transceiver. Figure 2 shows the major components of the CMU and the RPTs.

### 2.1.1   The Computer

The computer serves as the command center of the system. It displays the current sensor data from the RPTs and other information on the GUI. It also contains and executes the set of instructions and programs following the system process flow. All prior data collected from all RPTs are stored in the database in the computer and can be retrieved in Excel format.

The graphical user interface (GUI) in the system displays the current sensor data from all RPTs, and it is where the system settings are modified. Figure 3 shows the GUI used in the prototype that was developed using visual basic [6].

**Fig. 2** The monitoring system showing the major components inside the CMU and the RPT. The figure shows two RPTs only

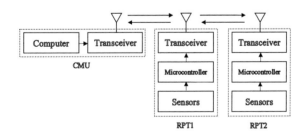

**Fig. 3** The graphical user interface of the prototype

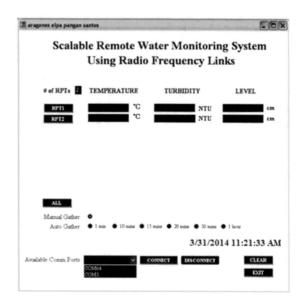

The proposed monitoring system uses a user-defined protocol developed by the authors for this particular application to make the system scalable and to avoid complications in the transmission and reception of data. At the CMU, it is used to command the RPTs to transmit their respective sensor data. Figure 4 shows the structure of the said CMU protocol.

The protocol starts with the header with a value of 33H (00110011B) in the prototype. This is the identifying information unique to the system. The destination indicates the destination RPT, e.g., 10H (00000010B), for RPT2. The source indicates the source of the data, in this case the CMU which is given 00H. The count is the number of RPTs deployed in the monitoring system. In the implementation of the system, the content of the checksum field is the sum of all bytes in the transmitted command data and is used in error detection.

The detailed process flow when the CMU sends the command data for the RPTs to send their sensor data is shown in Fig. 5a. In the chart are the values of the header (33H), and the destination (01H) means that the message is destined for RPT1 and the source (00H) is the default address of the CMU.

Figure 5b shows the process flow that the CMU follows when it is in the receive mode, particularly when receiving sensor data from RPT1. If the CMU receives sensor data from other RPTs, the information is discarded. The received data are also discarded if the transmission is from other wireless systems, that is, if the header is not 33H.

| Header | Destination | Source | Count | Checksum |
|--------|-------------|--------|-------|----------|
| 1 byte | 1 byte | 1 byte | 1 byte | 1 byte |

**Fig. 4** User-defined protocol for the command data emanating from the CMU

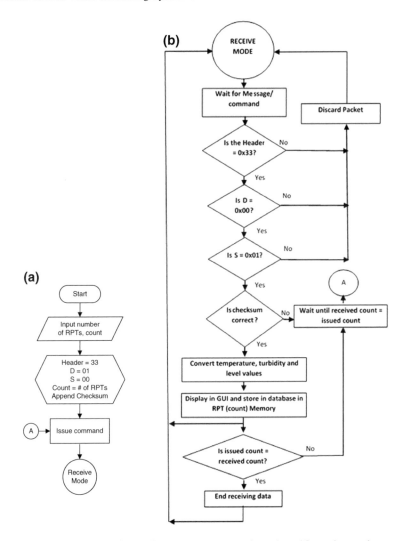

**Fig. 5** The flow diagrams that the CMU uses at **a** transmit mode and **b** receive mode

## 2.1.2 The CMU Transceiver

The commands generated by the computer are transmitted to their respective destination through the CMU transceiver. It is also its responsibility to receive the data from the nearest RPT for data processing. The transceiver at the CMU uses the Aerocomm, Inc. AC4790 transceiver module that operates at the 900 MHz ISM band. One unique feature of the transceiver module is its dynamic session extension and collision avoidance mechanism. This is done with the use of a proprietary scoring system to promote contention-free communication and to make sure that each node has fair access to the network.

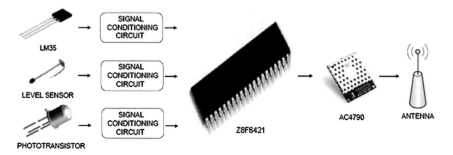

**Fig. 6** Diagram of remote process transceiver

## 2.2 The Remote Process Transceiver (RPT)

The RPTs in the monitoring system have the main function of gathering data from the sensors and send it to the CMU. Further, the first RPT will also serve as a repeater to retransmit the command data from the CMU to the second RPT and also can relay the data gathered by the second RPT to the CMU. An RPT is composed of a transceiver, a set of sensors, signal-conditioning circuits, and a microcontroller. Figure 6 shows all these components.

The proposed water monitoring system has three sensors: temperature, level, and turbidity. The system makes use of sensor components that are of low cost and locally available. The temperature sensor used in the prototype is the LM35D. It has an output of 10 mV/°C with a basic range of 0 °C through 100 °C. A conditioning circuit is designed such that the maximum output voltage would be +2 V.

The turbidity sensor involves the use of a light source and a photosensitive device. In the prototype, a light-emitting diode (LED) is used as the light source and a phototransistor (BPW77N) as the light detector. Lesser light is detected when the turbidity is high. Again, a signal-conditioning circuit in the form of an amplifier is used to produce a maximum output voltage of +2 V.

The water-level sensor uses a simple modified R-2R ladder circuit. The circuit implementation is affordable and easy to mount. The resistors are enclosed in a pipe before they are being submerged in water. In the implementation of the prototype of the proposed monitoring system, every 10-cm change in the water level produces a particular output voltage being produced by the R-2R ladder.

In each of the RPTs, a transceiver is used in the transmission and reception of data. Aerocomm, Inc. AC4790 transceiver assemblies are used, the same as the CMU transceiver. The specifications of the AC4790 transceiver can be read in [7]. Based on the experience, it is worth noting that the transmitter buffer on the Aerocomm device is 256 bytes. The device will begin RF transmission only once the buffer gets to that RF packet size. This can be accessed in the chip's EEPROM.

The microcontroller in the RPTs generally serves as an interface between the sensors and the transceiver located in each RPT. The microcontroller uses the Z8F6421 Z8 Encore microcontroller from Zilog. Specifically, the microcontroller

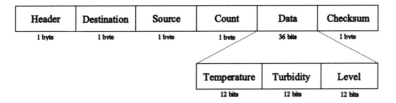

| Header | Destination | Source | Count | Data | Checksum |
|--------|-------------|--------|-------|------|----------|
| 1 byte | 1 byte | 1 byte | 1 byte | 36 bits | 1 byte |

| Temperature | Turbidity | Level |
|-------------|-----------|-------|
| 12 bits | 12 bits | 12 bits |

**Fig. 7** Format of the user-defined protocol for RPTs

performs the analog-to-digital conversion from the analog sensors. It also controls the transceiver through its UART port and performs the algorithm for the RPT. The sensor signals are applied at pins PB0, PB1, and PB2, while the transceiver is connected to it using its pins PD4, PD5, and PC0.

A user-defined protocol in the transmission of sensor data is used by the RPTs similar to the CMU protocol but with the inclusion of the 36-bit sensor data field. Figure 7 shows the format of the RPT protocol. The data field in the protocol contains the data from the temperature, turbidity, and level sensors, each with 12 bits.

Figure 8 shows the operation of the RPTs in the reception and transmission modes. Upon receiving the command, the header and the destination are identified whether they are correct; otherwise, the packet will be discarded. The correct value of the destination depends on the RPT.

When the source value is less than the RPT number, it denotes that the data are a command from the CMU or a relayed command from a nearer RPT. This will tend each receiving RPT to gather data from its sensors and then transmit. The last RPT will receive a count value of 01, and the relaying of command should cease. When the destination and header are correct and the source is greater than the RPT number, the received transmission is from an immediate distant RPT containing its sensor data. This should be relayed by the receiving RPT to the CMU or to the next nearer RPT.

## 3 Results and Discussion

This part of the paper presents the implementation and the results of the tests conducted to determine the performance of the proposed water monitoring system. The system was implemented using two RPTs and one CMU. Each RPT has its own set of sensors. The number of RPTs in the prototype is enough to demonstrate the features of the proposed system, including its scalability. Also, range tests were conducted to determine the effective distance between stations where reliable data transmission can still be achieved. Figure 9 shows photographs of the prototype of the proposed system.

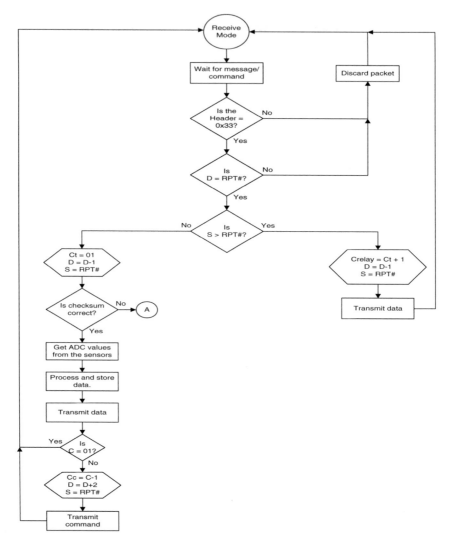

**Fig. 8** The RPT process flowchart

## 3.1 Sensor Tests

When conducting the sensor tests, the distance between the CMU and RPT1 and between the RPTs are well within the normal range of operation of the transceivers. The exact distance covered is 136 m each, which makes the overall distance equal to 272 m. Further, 15 different actual values of water temperature, turbidity, and depth are used. Three trials are conducted in each actual value. Table 1 shows the average sensor readings as compared to the actual values.

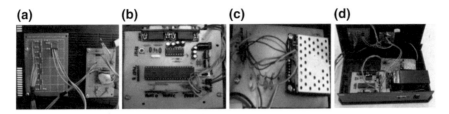

**Fig. 9** Photographs of the prototype. **a** Integration of sensors; **b** the Z8F6421 microcontroller with the associated components; **c** the AC4790 module transceiver assembly; **d** the transceiver components

**Table 1** Sensor test results sent by RPT to the CMU

| Sample | Temperature, °C | | Turbidity, NTU | | Level, cm | |
|---|---|---|---|---|---|---|
| | Actual value | Sensor value | Actual value | Sensor value | Actual value | Sensor value |
| 1 | 4 | 3.98 | 5 | 5.94 | 3.5 | 0 |
| 2 | 8 | 7.97 | 17 | 17.59 | 7 | 5 |
| 3 | 12 | 11.98 | 28 | 27.27 | 10.5 | 10 |
| 4 | 16 | 15.98 | 39 | 39.04 | 14 | 10 |
| 5 | 20 | 19.98 | 52 | 50.52 | 17.5 | 15 |
| 6 | 24 | 23.97 | 67 | 68.45 | 21 | 20 |
| 7 | 28 | 27.97 | 81 | 81.26 | 24.5 | 20 |
| 8 | 32 | 31.97 | 97 | 95.69 | 28 | 25 |
| 9 | 36 | 35.96 | 114 | 111.45 | 31.5 | 30 |
| 10 | 40 | 39.95 | 130 | 128.54 | 35 | 35 |
| 11 | 44 | 43.97 | 143 | 141.02 | 38.5 | 35 |
| 12 | 48 | 47.98 | 155 | 155.66 | 42 | 40 |
| 13 | 52 | 51.97 | 170 | 168.69 | 45.5 | 45 |
| 14 | 56 | 55.96 | 185 | 186.84 | 49 | 45 |
| 15 | 60 | 59.97 | 200 | 196.46 | 52.5 | 50 |

The accuracy of the data received at the CMU was tested by comparing the received sensor values at the CMU and the actual sensor values at RPT1 and RPT2. In the transmission of sensor data from the RPTs to the CMU, the received data at the CMU follow closely the readings of the sensors. The slight differences are due to the analog-to-digital conversion. All requests for sensor data by the CMU were replied with by the RPTs. The deviation of the CMU received data with the actual value is due mainly to the inaccuracies of the sensors.

**Table 2** Effective distance between CMU and the RPTs

| Distance (m) | | | CMU has received data | | | | | |
|---|---|---|---|---|---|---|---|---|
| Total | CMU and RPT1 | RPT1 and RPT2 | From RPT1 | | | From RPT2 | | |
| | | | Trial 1 | Trial 2 | Trial 3 | Trial 1 | Trial 2 | Trial 3 |
| 200 | 100 | 100 | Yes | Yes | Yes | Yes | Yes | Yes |
| 400 | 200 | 200 | Yes | Yes | Yes | Yes | Yes | Yes |
| 600 | 300 | 300 | Yes | Yes | Yes | Yes | Yes | Yes |
| 800 | 400 | 400 | Yes | Yes | Yes | Yes | Yes | Yes |
| 1000 | 500 | 500 | Yes | Yes | Yes | Yes | Yes | Yes |
| 1200 | 600 | 600 | Yes | Yes | Yes | Yes | Yes | Yes |
| 1400 | 700 | 700 | Yes | Yes | Yes | Yes | Yes | Yes |
| 1600 | 800 | 800 | Yes | Yes | Yes | Yes | Yes | Yes |
| 1800 | 900 | 900 | Yes | Yes | Yes | Yes | Yes | Yes |
| 2000 | 1000 | 1000 | Yes | Yes | Yes | Yes | Yes | Yes |
| 2200 | 1100 | 1100 | Yes | Yes | Yes | Yes | Yes | Yes |
| 2400 | 1200 | 1200 | Yes | Yes | Yes | Yes | Yes | Yes |
| 2600 | 1300 | 1300 | Yes | Yes | Yes | Yes | Yes | Yes |
| 2800 | 1400 | 1400 | Yes | No | No | No | No | No |
| 3000 | 1500 | 1500 | No | No | No | No | No | No |

## 3.2 Range Tests

The range tests were conducted to determine the maximum distance between stations where data transmission and reception are still possible. The distance between stations is varied every 100 m, and at each range value, transmission of command from the CMU and reception of sensor data from the RPTs are conducted. Table 2 shows the results of the range tests. Based on the results from three trials, the system as a whole is effective as a monitoring system up to about 1300 m between stations using the quarter-wave antenna supplied with the transceiver units. A much greater range can be achieved if higher-gain antennas are utilized.

## 4  Conclusions

The scalable remote water monitoring system using radio frequency links was proposed with the purpose of developing a monitoring system that is of low cost and utilizes locally available components and does not necessarily involve importing expensive parts. It serves as an immediate medium to monitor the fast degradation of the river systems in the Philippines. The communications part of the system used in the sending and reception of command and sensor data is effective up to 1300 m between stations. Issues such as addressing, error control, and data collision in wireless communications systems are managed with the use of the user-defined protocol. Further, the user-defined protocol utilized in the proposed system is effective in the implementation of the scalability feature. The scalability feature extends the coverage area of the monitoring system, allowing the number of RPTs to a maximum of 255 ($2^8 - 1$). However, the researchers recommend the use of industry-grade sensors, especially if the budget permits, to achieve more accurate readings. The use of directional antennas to increase the distance between stations is recommended. The researchers envision the use of the system for the transmission of data from sensors in various applications. It is similar but not limited to the systems described in [8, 9]. These applications may include habitat and ecosystem, seismic, civil structural health, groundwater contamination, industrial process, perimeter security and surveillance, and automated building climate control.

# References

1. Asian Development Bank (ADB) (August 2009) Country environmental analysis in Philippines. http://documents.worldbank.org/curated/en/2009/10/11377087/philippines-country-environmental-analysis
2. Asian Development Bank, Asia Pacific Forum (2007) Country paper Philippines. Asian Development Bank Outlook 2007. http://www.adb.org/publications/asian-water-development-outlook-2007
3. World Bank (December 2003) Philippine environment monitor 2003. http://documents.worldbank.org/curated/en/2003/12/3661476/philippines-environment-monitor-2003
4. Jiang P, Xia H, He Z, Wang Z (2009) Design of a water environment monitoring system based on wireless sensor networks. Sensors 9:6411–6434
5. Li X, Cheng X, Gong P, Yan K (2011) Design and implementation of a wireless sensor network-based remote water-level monitoring system. Sensors 11:1706–1720
6. Visual Basic Language (February 2008) https://msdn.microsoft.com/en-us/library/aa903378%28v=vs.71%29.aspx
7. AC4790 User Manual (2006) AC4790 900 MHz Transceivers Version 1.3. http://aerocomm.com/docs/users_Manual_AC4790.pdfd
8. Bitella G, Rossi R, Bochicchio R, Perniola M, Amato M (2014) A novel low-cost open-hardware platform for monitoring soil water content and multiple soil-air-vegetation parameters. Sensors 14:19639–19659
9. Palmer SCJ, Kutser T, Hunter PD (2015) Remote sensing of inland waters: challenges, progress and future directions. Remote Sens Environ 157:1–8

# Adaptive Packet Scheduling Scheme Based on Network Conditions in WLAN Using Fuzzy System

**Bakeel Maqhat, Mohd Dani Baba, Ruhani Ab Rahman, Anuar Saif, Md Mahfudz Md Zan, Mat Ikram, Mohd Asri Mansor and Farok Hj Azmat**

**Abstract** The IEEE 802.11n standard has introduced frame aggregation technique which can reduce the overhead and increases the channel utilization efficiency. However, the standard does not address the packet scheduling during the aggregation, and the aggregation scheme causes additional delays, particularly when waiting for other packets in the queue to construct the aggregated frame. Furthermore, the channel status should be addressed to improve the system performance. In this paper, we proposed a scheduling scheme called Dynamic Sensing Mechanism that handles the influence of network channel conditions during the transmission process. The mechanism decreases the number of expired packets during the retransmission by allowing the packets to be transmitted at an earlier time before their expiration time. The simulation results show an outstanding performance improvement for the proposed scheduling mechanism by reducing the packet loss ratio and increasing the system throughput.

**Keywords** 802.11n · Channel state · Fuzzy logic system · WLAN scheduling · QoS

## 1 Introduction

With the increasing demands for multimedia applications in wireless LAN systems, it has become essential to provide services with enhanced quality of service (QoS). The IEEE 802.11n standard is introduced to achieve more than 100 Mbps of throughput at the MAC layer and to enhance the QoS requirements [1].

B. Maqhat (✉) · M.D. Baba · R.A. Rahman · A. Saif · M.M.M. Zan · M. Ikram
M.A. Mansor · F.H. Azmat
Centre for Computer Engineering Studies, Faculty of Electrical Engineering,
Universiti Teknologi MARA, 40450 Shah Alam, Selangor, Malaysia
e-mail: bakeelyem@gmail.com

M.D. Baba
e-mail: mdani074@salam.uitm.edu.my

© Springer International Publishing Switzerland 2016
P.J. Soh et al. (eds.), *Theory and Applications of Applied Electromagnetics*,
Lecture Notes in Electrical Engineering 379, DOI 10.1007/978-3-319-30117-4_3

However IEEE 802.11 standard does not specify a scheduling algorithm, its left for the vendors to interpret. The current 802.11n scheduler inherits the priority mechanism of the legacy 802.11e Enhanced Distributed Channel Access (EDCA) scheduler. EDCA scheduler defines several access categories (ACs). A number of studies have been tried to implement scheduling mechanism for WLAN.

Wang and Zhuang [2] have proposed a novel token-based scheme to eliminate collisions and subsequently increases channel utilization. The authors have reported that by integrating voice and data traffic, the token-based scheduling can lead to a better result than DCF in terms of channel utilization. Inan et al. [3] introduced an application-aware adaptive HCCA scheduler for IEEE 802.11e WLANs. This scheduling algorithm is based on the Earliest Deadline First (EDF) scheduling discipline to make polling order based on the computed deadlines of the traffic. According to traffic specification along with instantaneous buffer occupancy information in time sensitive application, this algorithm schedules multimedia traffic by associating each QoS station with a distinct service interval (SI) and TXOP. However, the drawback is the additional hardware/firmware complexity that imposed by the algorithm which makes it not easy to implement. Frantti [4] introduces a fuzzy expert system to adapt the packet size for VoIP traffic in ad hoc networks. The model requires two variables input for the fuzzy system which are packet error rate and change of packet error rate. The results show that the expert system is capable to locate packet size values to the optimum level quickly along with the increment in the number of VoIP connections. Alsahag et al. [5] proposed a bandwidth allocation algorithm for the uplink traffic in mobile WiMAX called FADDR. The algorithm uses fuzzy logic control which is embedded in the scheduler with adaptive deadline-based scheme to guarantee a particular maximum latency for real-time traffics and maintain the minimum requirements for the non-real-time traffics. Seytnazarov and Kim [6] have introduced QoS-aware A-MPDU scheduler, which is applied to voice real-time traffic by controlling the delay time of whole buffering for A-MPDU and configure the buffering separately depending on access category and IP address for the destination. Ramaswamy et al. [7] have proposed Bi-Scheduler algorithm which separates frames depending on their access categories and schedules the VoIP traffic using A-MSDU aggregation, while scheduling the video and non-real-time traffic using the A-MPDU aggregation. Nevertheless, the algorithm is not effective under low-traffic load and the high sensitive traffic. Moreover, it may suffer from delay due to waiting for transmitting. The frame aggregation scheduler proposed by Selvam and Srikanth [8] have crystallized the approach of aggregate frames by calculating the deadline based on the earliest expiry time of a frame waiting in the queue and selecting dynamically the aggregation scheme based on frame aggregation size and bit error rate by using optimal frame size from the lookup table. However, this algorithm is restricted to deal with one type of traffic; thus, other traffic will suffer from delay and eventually may affect the QoS requirements of other traffics. The works in [9–11] have proposed schedulers by exploiting the A-MSDU attributes to enhance the system performance. Maqhat et al. [10] proposed a real time scheduling algorithm for A-MSDU aggregation (RSA-MSDU), which schedules the traffics that are time sensitive and have different lifetimes. The work requires a queue for the sender side, that is, the transmitting queue (TQ). In the TQ, the MSDUs will be sorted in ascending order based on their priorities. According to that, packet with the highest priorities will occupy the top of the queue. During aggregation, the packet is taken from the top of the TQ and attached with the aggregation headers. Then, the packet (subframe) is put into the aggregated frame which is referred as the superframe. The TQ sorting process is updated whenever a new packet is added to the TQ or an acknowledgment bitmap (ACK) is received. During the construction of the aggregation frame, only the packets that have the same destination address (DA) will be associated with the aggregation frame. The aggregated frame is transmitted as soon as it reaches the aggregation size limit or the aggregation delay limit ($T_{agg}$). The aggregation delay limit is estimated based on arrival time of the head packet in the queue ($H_{arv}$), the current time ($T_{crt}$), and the absolute time required to transmit an aggregated frame and receive its block acknowledgment ($T_{tx}$). The absolute time limit to transmit an aggregated frame is the $T_{agg}$ without compromising the delay of the aggregated packets. The $T_{agg}$ can be calculated by the following equation:

$$T_{agg} = (T_{crt} - H_{arv})/T_{tx} \qquad (1)$$

where $T_{agg}$ is the aggregation delay limit, $H_{arv}$ is the arrival time of the head packet in the queue, and $T_{crt}$ is the current time.

The TQ is updated once a Block Acknowledgment is received. When packet is received correctly, it will be removed from the TQ; otherwise, it will be considered as corrupted. The corrupted packet will gain a high sending priority by placing it at the top of TQ and will be retransmitted at the head of the next aggregated super-frame. The drawback of this scheme is that sending of the superframe is done based on the time required for transmitting the superframe, without taking into consideration the lifetime of the retransmitted packets. Ignoring such factor will result in removing the dead packets which in turns affect the packet loss ratio.

Addressed this issue is presented in this paper by introducing a Dynamic Sensing Mechanism (DSM) scheme, which handles the influence of network channel conditions for the transmission process. DSM scheme uses the fuzzy expert system to calculate dynamically the transmission time.

## 2 Dynamic Sensing Mechanism

Dynamic Sensing Mechanism (DSM) scheme handles the influence of network channel conditions for the transmission process. This is done by investigating the factors that affect the network channel state such as the noise and network load. DSM scheme estimates the traffic load and noise and calculates the time required for successful transmission. The estimation in this mechanism is based on the assumption that by increasing the traffic load and noise will increase the number of failed packets and thus will increase the time required for successful transmission. Hence, the DSM scheme will allow the retransmission attempts before packets expired. The mechanism relies on two sensors, erroneous channel sensor (ECS) and traffic load sensor (TLS), where ECS indicates the amount of failed packets, caused by the noise. TLS indicates the attempts of superframes transmitted which caused by traffic congestion. Moreover, the noise has an effect on the traffic load. Having a high noise in the network causes a bit error in the packet transmission and increases the retransmission attempts. DSM mechanism is able to decrease the number of expired packets during retransmission by allowing the packets to be transmitted at an earlier time before expiration. This behavior will enhance the performance by reducing the packet loss ratio.

To investigate the effect of the network load and noise, we have conducted some experiments in different channel conditions. In these experiments, the number of users varies from 2 to 60 and the channel condition varies from error-free channel to erroneous channel. The results are collected and then analyzed statistically using SPSS program. During the analysis, we used the linear regression between ECS, TLS, traffic load, and noise. We set the noise and traffic load as independent factors and ECS and TLS as dependent factors during the analysis. Results show that traffic load and noise have a significant effect on TLS. Noise and traffic load predict 0.911 of TLS, (0.654 and 0.254) for the two factors, respectively. However, ECS is affected significantly only by noise and not by traffic load. Noise predicts 0.988 of ECS, whereas traffic load predicts only 0.008 that is a negligible value. The linear regression results give an indication to what extent the noise and load affect ECS and TLS. Therefore, we can rely on ECS and TLS in determining the performance of the network.

The DSM allows the packets to be transmitted even in bad network conditions. In this case, the lifetime of the packets will be the criteria for removing the packet from transmitting queue rather than the number of retry attempts.

## *2.1 Calculate the Transmission Time Using Fuzzy System*

Enhance the work by performing adaptive and dynamic mechanism which computes an optimal time for the transmission process better than in static methods.

DSM uses an expert system based on fuzzy logic to enable the scheduler to make a transmission decision based on the network channel conditions. This can be implemented with an embedded system. Fuzzy logic is a concept which helps computers in making decisions in a way which resembles human behaviors.

DSM scheme dynamically updates the weight of which is the amount of time can be added to the aggregation delay limit to allowed packets to be transmitted before expiration. This weight is an embedded fuzzy system output. We aim in this work to ensure that we compute an optimal time for frame transmission decision in each transmission process which take into account packet deadline and channel status. Figure 1 describes the proposed DSM with embedded fuzzy system.

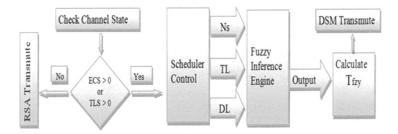

**Fig. 1** DSM scheme using embedded fuzzy system

This work develops an embedded fuzzy system to dynamically compute the required time for transmission decision more accurately and with low complexity.

The sensors provide input variables into the system from which decisions can be made, three variables used as the embedded fuzzy system inputs. First is the amount of noise (Ns), the second variable is amount of traffic load (TL), and the third is the head-of-line packet deadline (DL).

## 2.2 Fuzzy Reasoning Inference Engine Control Model

As mentioned earlier, the embedded fuzzy system uses three variables and one variable as inputs and output, respectively. In this system, we design fuzzy logic from fuzzy set theory to design a theoretical structure for the linguistic information where the most important design is to utilize the expert information for the rule base creation. An individual based inference method with Mamdani's design [12] is used, where the inference system rules are jointed into one value. The fuzzy system consists of the following stages: fuzzification, fuzzy reasoning inference, and defuzzification.

The role is to dynamically analyze all input traffics and combine them into one overall fuzzy set. Firstly, the fuzzification process handles three input variables: Ns, TL, and DL, for the overall system. Then, reasoning inference mechanism contains the rule base to manipulate the input variables as shown in Fig. 2. At this point, the actual decision is made representing the human expert process which performs to the linguistic behavior to obtain the output value.

Lastly, the defuzzification phase calculates crisp numerical values to obtain the required weight, which provides an indication of the priority for the scheduler.

Three linguistic levels have been defined for input variables: Ns, TL, and DL, namely low, medium, and high, and five levels for output such as very low, low, medium, high, and very high. As the utilized fuzzy system considers three variables as inputs and three membership functions are considered for each, subsequently the rule base composed of 27 rules (see Table 1). The dynamically normalized scale for the DL input variable is formed from 0 to 200 ms, whereas the dynamically normalized scale for the other inputs and output variables is formed from 0 to 1.

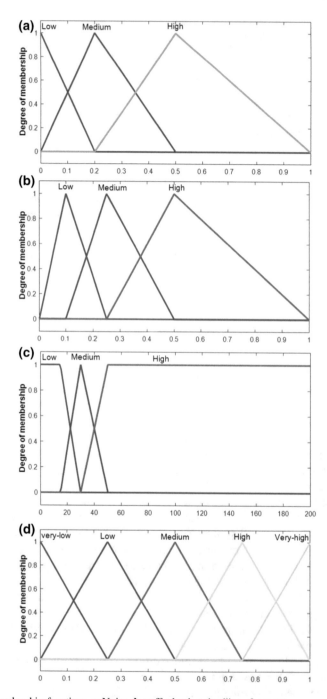

**Fig. 2** Membership functions. **a** Noise, **b** traffic load, **c** deadline, **d** output $w_{tf}$

| **Table 1** Fuzzy system rule base | Rules N | Noise | T. Load | Deadline | Output $w_{tf}$ |
|---|---|---|---|---|---|
| | 0 | Low | Low | Low | Low |
| | 1 | Low | Medium | Low | Low |
| | 2 | Low | High | Low | High |
| | 3 | Low | Low | Medium | Very low |
| | 4 | Low | Medium | Medium | Low |
| | 5 | Low | High | Medium | Medium |
| | 6 | Low | Low | High | Very low |
| | 7 | Low | Medium | High | Very low |
| | 8 | Low | High | High | Low |
| | 9 | Medium | Low | Low | High |
| | 10 | Medium | Medium | Low | Very high |
| | 11 | Medium | High | Low | Very high |
| | 12 | Medium | Low | Medium | Medium |
| | 13 | Medium | Medium | Medium | Medium |
| | 14 | Medium | High | Medium | Medium |
| | 15 | Medium | Low | High | Low |
| | 16 | Medium | Medium | High | Low |
| | 17 | Medium | High | High | Medium |
| | 18 | High | Low | Low | Very high |
| | 19 | High | Medium | Low | Very high |
| | 20 | High | High | Low | Very high |
| | 21 | High | Low | Medium | Medium |
| | 22 | High | Medium | Medium | Medium |
| | 23 | High | High | Medium | High |
| | 24 | High | Low | High | Medium |
| | 25 | High | Medium | High | Medium |
| | 26 | High | High | High | Medium |

Weight value ($w_{tf}$) obtained from the fuzzy inference control system is the time factor ratio which is used to compute the optimal time that enables the scheduler to make a decision to send the superframe based on the network channel conditions in order to support packets retransmission. The new aggregation delay limit is given by Eqs. 2 and 3.

$$T_{fzy} = T_{tx} + DL \times w_{tf} \qquad (2)$$

$$T_{agg} = \frac{(T_{crt} - H_{arv})}{T_{fzy}} \qquad (3)$$

where $T_{fzy}$ is the new transmission time which is computed by fuzzy system. This scheme is capable of decreasing the number of expired packets during retransmission by allowing the packets to be transmitted at an appropriate earlier time before expiration, particularly in the bad channel conditions. This behavior will enhance the QoS requirements by reducing the packet loss ratio and increase the system throughput. The pseudocode of DSM system is presented in Algorithm 1.

## 2.3 Explain Fuzzy System Work by Examples

The designed fuzzy system will update the weights of the transmission time of IEEE 802.11n scheduler. The primary aim of the system was to adapt itself to variations of network channel conditions by providing an optimal time to make frame transmission decision to allow the packets to be transmitted before expiration.

In this section, as an example to explain this system: suppose the sensors sense the amount of noise and traffic load with the values of 0.27 and 0.14 for noise and traffic load, respectively. And the scheduler specifies the deadline with the value of 19 ms. The fuzzy system will read these variables as its inputs. Considering the membership functions in Fig. 2, we can make out the linguistic values for three input variables of 0.27, 0.14, and 19 and can be read as follows:

Let the amount of noise be 0.27 which after fuzzification is in linguistic form medium at grade membership of 0.8 and high at grade membership of 0.20 (see Fig. 2a). Let the load be 0.14 which is after fuzzification low at grade membership of 0.7 and medium at grade membership of 0.3 (see, Fig. 2b). In addition, the deadline (DL) is 0.19 ms which means it is low at grade membership of 0.65 and medium at grade membership of 0.35 (see, Fig. 2c).

Applying fuzzy reasoning with rule base from Table 1 and Fig. 2, we can read as:

- If the noise is medium (0.8) AND load is low (0.70) AND DL is low (0.65) THEN the weight value ($w_{tf}$) is high at grade membership of 0.73 [Rule 9].
- If the noise is medium (0.80) AND load is low (0.70) AND DL is medium (0.35) THEN $w_{tf}$ is medium at grade membership of 0.26 [Rule 12].
- If the noise is medium (0.80) AND load is medium (0.30) AND DL is low (0.65) THEN $w_{tf}$ is very high at grade membership of 0.26 [Rule 10].
- If the noise is medium (0.80) AND load is medium (0.30) AND DL is medium (0.35) THEN $w_{tf}$ is medium at grade membership of 0.26 [Rule 13].
- If the noise is high (0.2) AND load is low (0.70) AND DL is low (0.65) THEN $w_{tf}$ is very high at grade membership of 0.23 [Rule 18].
- If the noise is high (0.2) AND load is low (0.70) AND DL is medium (0.35) THEN $w_{tf}$ is medium at grade membership of 0.23 [Rule 21].
- If the noise is high (0.2) AND load is medium (0.30) AND DL is low (0.65) THEN $w_{tf}$ is very high at grade membership of 0.23 [Rule 19].
- If the noise is high (0.2) AND load is medium (0.30) AND DL is medium (0.35) THEN $w_{tf}$. is medium at grade membership of 0.23 [Rule 22].

From the above rules and using Mamdani's inference, we can conclude that the weight value will be high at grade membership of 0.73 and medium at grade

**Table 2** Examples for input parameters and their fuzzified values

| Inputs variables | | | Crisp output $w_{tf}$ |
|---|---|---|---|
| Noise | T. load | Deadline | |
| 0.27 | 0.14 | 19 | 0.72 |
| 0.04 | 0.60 | 8 | 0.80 |
| 0.20 | 0.10 | 50 | 0.25 |
| 0.60 | 0.45 | 36 | 0.625 |
| 0.80 | 0.90 | 60 | 0.50 |
| 0.05 | 0.26 | 100 | 0.085 |

membership of 0.27. The final value of output weight of time is calculated after defuzzification which gives a crisp value of 0.72. It will be used as time factor to calculate an optimal time for scheduler to transmit a superframe with taking into consideration the bad channel conditions.

For every three input variables, the intelligent fuzzy system generates one crisp value. More examples are shown in Table 2.

---

**Algorithm 1.** Pseudo code of DSM algorithm

```
1:    if superframe is not null  then
2:        Check the channel state
3:        if attempts of superframes > 0  then
4:            TLS=attempts of superframes/Max(attempts of superframes)
5:        end if
6:        if retrylimit > 0  then
7:            ECS = failed  packets /(total  packets )
8:            fuzzy first input= Ns
9:            fuzzy second input=TL
10:            fuzzy third input=DL
11:        end if
12:    calculate transmission time Tfzy Eq (2)
13: end if
13: if channel is idle l then
14:        send the superframe
15: end if
```

---

# 3   Results and Discussions

We have carried out several simulation experiments using the network simulator (NS2) in order to evaluate the scheduling of time sensitive traffics in terms of packet loss, throughput, and average delay. Moreover, we have used the simulation scenarios number 17 of the point-to-point usage model [13]. The scenario consists of a single-hop WLAN in which the transmission power of all the high-throughput STAs is high enough to ensure no hidden terminals in the network. All the stations are operating over a 20 MHz. Furthermore, the network traffic is composed of VoIP with a packet size of 120 bytes, video conferencing with a packet size of 512 bytes, and Internet video/audio streaming with a packet size of 512 bytes. The VoIP rate is 0.96 Mbps, video conferencing rate is 2 Mbps, and the Internet streaming video/audio rate is 2 Mbps. The characteristic of traffic is given in Table 3.

The data rates are set to 150 and 300 Mbps and basic rate is set to 54 Mbps. The average delay, throughput, and packet loss ratio of the different traffic are examined under a different number of stations (i.e., varied from 10 to 60) and different data rates with a bit error rate (BER) is set to $10^{-4}$. Other simulation parameters are listed in Table 4.

The average delay in the DSM scheme and RSA-MSDU is shown in Figs. 3, 4, and 5. The proposed DSM scheme achieves smaller average delay due to its ability to transmit the packets at a sufficient time before their expiration time; thus, the

**Table 3** Application characteristics

| Application | Priority | Data rate (Mbps) | Lifetime (ms) |
|---|---|---|---|
| VoIP | Very high | 0.096 | 30 |
| Video conferencing | High | 2 | 100 |
| Internet streaming | Medium | 2 | 200 |

**Table 4** Simulation parameters

| Parameter | Value |
|---|---|
| $T_{SIFS}$ | 16 μs |
| $T_{PHYhdr}$ | 20 μs |
| $T_{idle}$ | 9 μs |
| $CW_{min}$ | 16 |
| $T_{DIFS}$ | 34 μs |
| Basic rate | 54 Mbps |

**Fig. 3** VoIP average delay

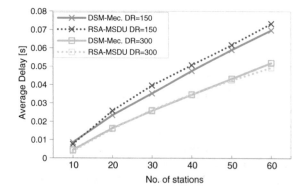

**Fig. 4** Video conferencing average delay

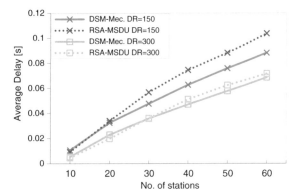

**Fig. 5** Internet streaming video/audio average delay

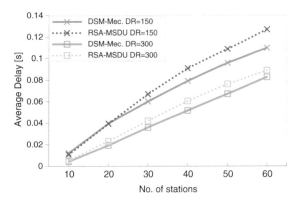

packets will not suffer a long queuing delay. The lower the contending station, the smaller the average delay as the stations will have frequent access to the medium and the packets will not suffering a long queuing delay. The performance gain for the DSM scheme over the RSA-MSDU is about 6 % for VoIP in the case of 150 Mbps and no gain is recorded under 300 Mbps data rate where packets will not suffer from long queuing delay due to the high data rate. For video conferencing, the gain is about 14 % in the case of 150 Mbps, while it is about 6 % at high data rates. Even with the high delay tolerance in video conferencing, DSM still able to improve the performance. The DSM enhancement for Internet streaming video/audio is about 12 % for both data rates.

The packet loss ratio of DSM mechanism compared to RSA-MSDU is shown in Figs. 6, 7, and 8. With a small number of competing network stations, the packet loss will be small because the network will have a low number of superframes compete to transmit. The packet loss will rise with the increasing number of stations. Moreover, the figures show the outstanding performance of our mechanism in reducing the packet loss ratio at the high noise and the high traffic load. Figure 6 shows an enhancement in VoIP packet loss of about 30 % for both data rates, while Fig. 7 shows the video conferencing enhancement which is about 52 and 65 % for both data rates.

**Fig. 6** VoIP packet loss ratio

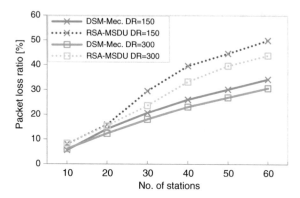

**Fig. 7** Video conferencing
packet loss ratio

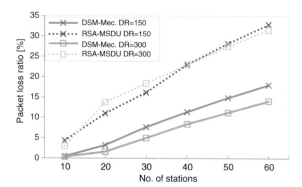

**Fig. 8** Internet streaming
video/audio packet loss ratio

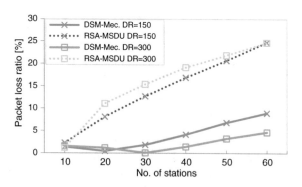

Internet streaming video/audio follows the same behavior as the other traffics
and scores an enhancement of 72 and 87 % under 150 and 300 Mbps, respectively,
see Fig. 8. The previous results match with our assumption that the DSM mechanism will have a great impact on reducing the packet loss. Moreover, the results
show that whenever the delay bound of the traffic is high, the better the DSM

performance. The packet loss enhancement is increased whenever increased the lifetime of the traffics.

Figures 9, 10, and 11 show the throughput performance of the proposed DSM scheduler scheme under a different number of stations. In a large number of stations, the collisions occur repeatedly and impact the system performance. The system throughput keeps decreasing while the network stations increase. Nevertheless, the system throughput of the DSM mechanism reaches about 38 and 53 Mbps, whereas the RSA-MSDU approaches 32 and 47 Mbps at high traffic load under 150 and 300 Mbps, respectively, see Fig. 9. The throughput gain reaches more than 8 % for both video conferencing and Internet streaming video/audio under 150 Mbps. In a data rate of 300 Mbps, the throughput gain reaches 5 and 6 % for video conferencing and Internet streaming video/audio, respectively, see Figs. 10 and 11. The increase in the throughput is attributed to the enhancement of the system efficiency, due to reducing the amount of loss.

**Fig. 9** System throughput

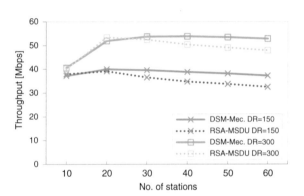

**Fig. 10** Video conferencing throughput

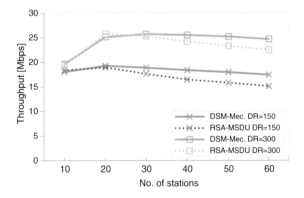

**Fig. 11** Internet streaming video/audio throughput

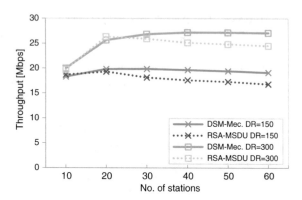

## 4 Conclusion

In this work, we have introduced a scheduling mechanism called DSM scheme which handles the influence of network channel conditions for the transmission process by estimating the traffic load and noise and then calculates the required time for successful transmission. DSM mechanism uses two sensors: ECS and TLS to sense the amount of noise and traffic congestion in the network. Accordingly, DSM scheme employs the fuzzy expert system to dynamically compute the superframe time transmission. The simulation results show that DSM scheme can significantly improve the system performance by reducing the packet loss ratio to about 80 % and increasing the system throughput.

**Acknowledgments** The authors would like to acknowledge the Research Management Institute of Universiti Teknologi MARA for providing the research grant for the Project No: 600-RMI/DANA 5/3/REI (2/2015).

## References

1. Xiao Y (2005) Performance analysis of priority schemes for IEEE 802.11 and IEEE 802.11 e wireless LANs. IEEE Trans Wireless Commun 4(4):1506–1515
2. Wang P, Zhuang W (2008) A token-based scheduling scheme for WLANs supporting voice/data traffic and its performance analysis. IEEE Trans Wireless Commun 7(5):1708–1718
3. Inan I, Keceli F, Ayanoglu E An adaptive multimedia QoS scheduler for 802.11 e wireless LANs, pp 5263–5270
4. Frantti T Fuzzy packet size optimization for delay sensitive traffic in ad hoc networks, pp 2633–2637
5. Alsahag AM, Ali BM, Noordin NK, Mohamad H (2014) Fair uplink bandwidth allocation and latency guarantee for mobile WiMAX using fuzzy adaptive deficit round robin. J Netw Comput Appl 39:17–25
6. Seytnazarov S, Kim Y-T QoS-aware MPDU aggregation of IEEE 802.11 n WLANs for VoIP services, pp 2–4

7. Ramaswamy V, Sivarasu A, Sridharan B, Venkatesh H (2014) A bi-scheduler algorithm for frame aggregation in IEEE 802.11 n. arXiv preprint arXiv:1401.2056
8. Selvam T, Srikanth S A frame aggregation scheduler for IEEE 802.11 n, pp 1–5
9. Maqhat B, Baba MD, Rahman RA A-MSDU real time traffic scheduler for IEEE802. 11n WLANs, pp 286–290
10. Maqhat B, Baba MD, Ab Rahman R, Saif A (2014) Scheduler algorithm for IEEE802. 11n wireless LANs. Int J Future Comput Commun 3(4):222–226
11. Maqhat B, Dani Baba M, Rahman RA, Saif A Performance analysis of fair scheduler for A-MSDU aggregation in IEEE802. 11n wireless networks, pp 60–65
12. King PJ, Mamdani EH (1977) The application of fuzzy control systems to industrial processes. Automatica 13(3):235–242
13. Stephens AP et al (2004) Technical report, IEEE 802.11n working document 802.11-03/802r23

# Designing and Experimental Validation of an Electromagnetic Setup at MHz Frequency Under Different Propagation Media

**Muhammad Adil, Hasnah Mohd Zaid, Lee Kean Chuan
and Noor Rasyada Ahmad Latiff**

**Abstract** For successful underwater electromagnetic (EM) wave operation, knowledge is required of the wave propagation properties of seawater overall distances both short and long. This paper presents a novel application of EM wave propagation through seawater, in the field of nano-EOR by manipulating the nanoparticle characteristics under EM waves. For this purpose, a solenoid-based 167 MHz transmitter is designed, simulated, and fabricated in the laboratory and has been studied in two steps under air, tap water, and salt water as a transmission medium. Firstly, a laboratory-scale model was simulated by using computer simulation technology (CST) to test the designed solenoid under different background media, whereas in second step, these simulation results were validated experimentally using a scale tank with a scale factor of 4360. It is noted that the solenoid has been best to propagate at an optimum applied voltage of 1.5 V in salt water. The comparative results at 1.5 V show that the magnitude of the EM waves was found to deviate up to 8.23 % in salt water, while it deviates by 21.3 and 44.3 % for tap water and air, respectively. This proved its suitability to employ in underwater EOR application.

**Keywords** Electromagnetic wave propagation · Simulation modeling · Solenoid · EM shielding · Attenuation

M. Adil (✉) · H.M. Zaid · L.K. Chuan · N.R.A. Latiff
Fundamental and Applied Science Department, Universiti Teknologi PETRONAS,
Bandar Seri Iskandar, 32160 Tronoh, Perak, Malaysia
e-mail: muhammadadil86@hotmail.com

H.M. Zaid
e-mail: hasnamz@petronas.com.my

L.K. Chuan
e-mail: lee.kc@petronas.com.my

N.R.A. Latiff
e-mail: syasya.latiff@gmail.com

© Springer International Publishing Switzerland 2016
P.J. Soh et al. (eds.), *Theory and Applications of Applied Electromagnetics*,
Lecture Notes in Electrical Engineering 379, DOI 10.1007/978-3-319-30117-4_4

# 1  Introduction

Electrorheological (ER) fluids are a type of nanofluid whose rheological characteristics can be changed upon the application of an electric field [1]. Such rheological variation is normally reversible and occurs within 10 ms. Hence, ER fluids
are occasionally denoted as a type of smart fluid. The diverse application potential
has made ER fluids a persistent topic of study ever since their discovery over six
decades ago by Winslow [2]. There has been extensive research into the basic
mechanism of the ER effect [3]. The latest discovery of the giant electrorheological
(GER) effect [4], together with its appeal in the basic science of nanoparticles and
their dynamics, has provided a new direction in this area.

However, our primary study focuses on using ER fluids in upstream oil industries, especially for enhanced oil recovery (EOR) purposes. Haroun et al. [5] has
shown that certain nanoparticles, such as copper oxide and nickel oxide (50 nm),
can be used to enhance oil recovery in low permeability carbonates (77–149 md,
porosity between 12 and 24 %) from Abu Dhabi using an applied electrical field.
The authors termed it electrical EOR (EEOR). As a curiosum, it has recently been
shown experimentally that oil droplets can be deformed when surrounded by
nanoparticles under the influence of an external electrical field [6]. It speculates that
this process may in part be responsible for the EEOR-induced improved oil
recovery. It is clearly shown from the literature above that there are not much
technical evaluations and documentations found on the rheological behavior for this
type of fluids under electromagnetic waves (EM), with respect to EOR. This forms
the primary motivation of this work, which deals with the development of novel
way of measuring the viscosity of nanoparticles under the presence of EM waves.
A simple schematic setup is shown in Fig. 1.

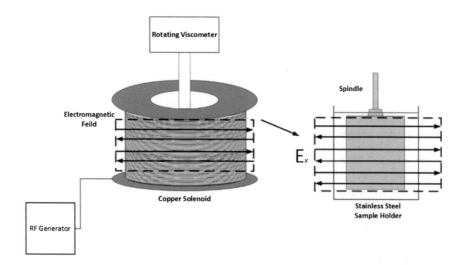

**Fig. 1** Simple schematic of EM experimental setup for viscosity measurement of nanofluid

When nanofluids are activated by an applied electric field, it behaves as a Bingham plastic (non-Newtonian fluid), where the nanofluids shear as a fluid once the yield point has been reached. Hence, the mobility of the fluid can be efficiently controlled by regulating the applied field. This paper describes the designing and experimental aspects of the electromagnetic system and implies to investigate the change in viscosity of nanofluids. The computational designing is carried out by computer simulation technology (CST) package, which involves the design and simulation of solenoid in three different environments including air, tap water, and salt water (≈seawater concentration). This was done to evaluate the electric field strength with respect to the laboratory-scale frequency and propagation medium, in order to optimize the voltage (≈applied field strength) needed to be supplied for activating the nanoparticles to achieve ER effect. Subsequently, the simulation results were comparatively evaluated by experimental observations, done in a laboratory-scale tank with 4360 scale factor under the similar environment.

## 2  Methodology

### 2.1  Laboratory-Scale Calculation

The Baronia field has been chosen for a reference, having well spacing of 1000 ft. Based on the well spacing, the scale factor calculation (Table 1) has been done to determine the optimum frequency to perform the CST simulations and the corresponding experiments. Full-scale ($d_{fs}$) and the laboratory-scale ($d_{lab}$) dimension ratios are related by $n$ scale factor as by [7],

$$n = \frac{d_{fs}}{d_{lab}} \tag{1}$$

For the frequency ($f$) and wavelength ($\lambda$),

$$f_{lab} = n^2 f_{fs} \tag{2}$$

$$\lambda = 2\pi \sqrt{\frac{2}{\mu\sigma\omega}} \tag{3}$$

**Table 1** Laboratory-scale calculation at scale factor of 4360 for seawater, as a reference case

| dfs (ft) | Field scale frequency $f_{fs}$ (Hz) | Laboratory-scale frequency $f_{lab}$ (MHz) | Laboratory-scale wavelength $\lambda$ (m) |
|---|---|---|---|
| 1000 | 8.8 | 167.3 | 0.07 |

## 2.2 Simulation Modeling

The simulation protocols related to the study data acquisition are explained in this section. Computer simulation technology (CST) software uses Maxwell equations was introduced to design and solve the different models with finite element method. In this paper, CST EM STUDIO had been used to simulate a laboratory-scale model by using a conventional copper solenoid with and without the stainless steel (SS304) sample holder as depicted in Fig. 1. The solenoid was specifically designed to facilitate the generation of EM waves with low losses at a scale-down frequency of 167 MHz. The detailed schematic of solenoid is presented in Fig. 2a. CST modeling have a potential to investigate the effect of electromagnetic fields (*E*, *D*, *B*, and *H*) with respect to varied transmitting frequencies and voltage, distance between the source and the different traveling medium, etc.

Certain steps are involved in order to generate the CST simulation mode. This investigation uses the parameters for modeling inspired by laboratory-scale calculation. Arranging background parameters is the first step that consists of setting model area of 14 × 4 cm, which could be easy to replicate experimentally. Simulation models were classified as models A, B, and C with respect to propagation medium of air, tap water, and salt water, respectively. These models are illustrated in Fig. 3. For model A, the propagation medium was only air having a thickness of 20 cm. For model B, tap water depth of 20 cm was chosen, with an air thickness of 20 cm over it, whereas in the case of model C, tap water layer was replaced with salt water. All the layers are allotted with their respective conductivities, permittivities, and permeability values as tabulated in Table 2.

Second step was to set parameters for copper solenoid as EM transmitter. In our case, frequency of 167 MHz and a voltage varying between 1 and 2 V was

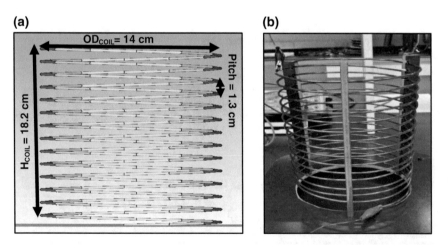

**Fig. 2 a** Dimensions of solenoid at the designed frequency of 167 MHz, while **b** fabricated solenoid for experimental validation

**(a)**            **(b)**            **(c)**

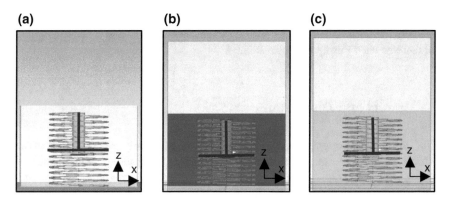

**Fig. 3** Simulation models used to generate the E-field strength data with a background as **a** air (Model A), **b** tap water (Model B), and **c** salt water (Model C)

**Table 2** Values assigned to the physical properties of propagation media

| Medium | Electric conductivity $\sigma$ (S m$^{-1}$) | Material density $\rho$ (kg m$^{-3}$) | Relative permittivity $\varepsilon_r$ | Relative permeability $\mu_r$ |
|---|---|---|---|---|
| Air | $1 \times 10^{-11}$ | 1.293 | 1.0006 | 1.0 |
| Tap water | 1.59 | 1000 | 78 | 0.991 |
| Salt water | 3.53 | 1025 | 74 | 0.991 |

set along $x$ direction, with horizontal and vertical receivers array as shown in Fig. 3. Subsequently, electric boundary conditions were applied and then run a full wave solver to initiate the simulation.

## 2.3 Experimental Setup/Validation

The next step was to undertake experimental trails in a laboratory test tank, to replicate the simulation models. The testing tank, consists of a transmitter and a receiver, was constructed from plastic and had dimensions of 3.55 m (W), 3.10 m (H), and 4.55 m (L). The transmitter (i.e., 10 AWG copper solenoid, Fig. 2b) was placed at the bottom of the test tank and was fully submerged. While a RF generator (Agilent E4421B) was used to feed the transmitter at a fixed frequency of 167 MHz and variable voltage between 1 and 2 V. On the other hand, the receiver was a coated single-loop antenna directly connected to the spectrum analyzer (Anritsu MS2036A). The receiver was handheld in the tank allowing the receiver to record the field strength at varied distance to replicate the results from CST model. The electric field strength between the transmitter and the receiver was initially measured as a function of distance within the laboratory test tank, before the stainless steel sample holder was mounted in the center of solenoid to measure the field strength within it. The receiver with a diameter of 25 mm and a fixed length of 10 cm was used during the measurement.

# 3 Results and Discussion

This study is thoroughly based on the response of merely $x$ component of the electric field. Therefore, $E_x$ component response is compared between simulated and experimental results, in order to observe the change in strength of electric field with varying background media. The detailed $E_x$ data have not been presented here, but only the peak strength of E-field for every applied voltage is presented.

## 3.1 E-field in Air

The spectral response of maximum E-field in air is shown in Fig. 4 at a frequency of 167 MHz. According to simulation results, the peak value for E-field is found to be 177, 266, and 355 mV at a supplied voltage of 1, 1.5, and 2 V. However, these values show a huge reduction in 1.56, 2.11, and 2.025 mV when compared to the experimental measurements in air. This big deviation could be the reason of confined boundary condition for simulated model, while experimental results present a big loss in the field strength due to infinite boundary.

Similar behavior is observed in the presence of stainless steel chamber (Fig. 5), for both simulated and experimental results where the E-field strength reduces due to the shielding effect. The percentage deviation is found to be 43.5, 44.3, and

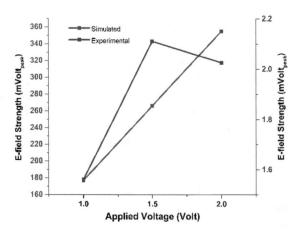

**Fig. 4** Maximum E-field strength in air via simulation and experiment without SS chamber

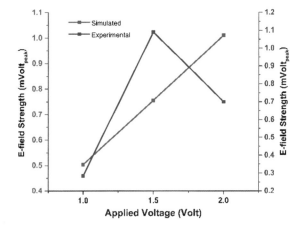

**Fig. 5** Comparative results of E-field strength at peak with SS chamber in air (model A)

30.6 % for 1, 1.5, and 2 V, respectively. It is further concluded that the experimental values are more acceptable as air is naturally a bad transmission medium due to poor conductivity ($\sigma/\omega\varepsilon \ll 1$).

## 3.2 E-field in Tap Water

The CST simulation and experimental results are compared for the 167 MHz-designed solenoid in tap water in Fig. 6. The received signal strength is measured to be 54.1, 81.1, and 108 mV (computationally) to 26.97, 43.92, and 42.8 mV (experimentally), at the vertical distance of 9 cm.

The measurement depth is limited to a maximum of 10 cm, equal to the total length of receiver. However, this limitation does not raise any question to the

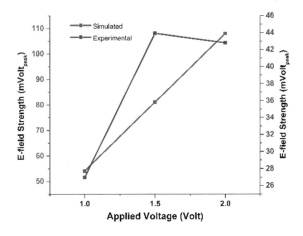

**Fig. 6** Simulation results and their experimental validation in tap water at an applied voltage of 1, 1.5, and 2 V

**Fig. 7** Peak value of E-field evaluated for model B in SS chamber at an applied voltage of 1, 1.5, and 2 V

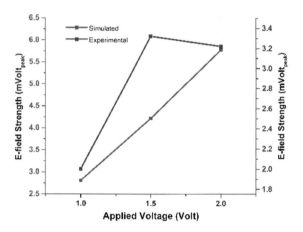

credibility of experimental results as the focus point of the EM field is at the center of solenoid corresponding to maximum E-field strength. Shielding effect has again played an important role to reduce the propagation of EM field in water by absorbing the energy of the signal, in the presence of SS sample holder (Fig. 7). For model B, the electric field strength is peaked at 2.81, 4.22, and 5.78 mV inside the SS chamber, compared to experimental values of 2, 3.32, and 3.22 mV with a difference of 28.8, 21.3, and 44.2 %, respectively. This can be explained by the fact that tap water is highly conductive and possesses a low attenuation due to low salinity at the radio frequency [8].

## 3.3   E-field in Salt Water

The comparative results of EM propagation through salt water (≈seawater concentration) without SS chamber are shown in Fig. 8. In salt water, the simulation presented a sharp peak at a frequency of 167 MHz with a very small E-field magnitude of 0.0298, 0.0447, and 0.0597 mV at an applied voltage of 1, 1.5, and 2 V, respectively, while the corresponding experimental values are found to be 0.78, 1.12, and 1.11 mV. There are two main factors affecting the EM propagation in salt water: First, because of the high value of skin depth, propagation of EM waves is limited and the wave amplitude remarkably decays along propagation [9]; secondly, due to the high salinity, which promotes the attenuation by spreading of energy as it propagates away from the transmitter.

In the presence of SS chamber (Fig. 9), the magnitude of E-field strength shows a surprisingly noticeable increment which is verified by both simulation and experimental results. For model C, the E-field value was topped at 1.78, 2.67, and 3.56 mV in comparison with 1.75, 2.45, and 2.35 mV evaluated experimentally. The percentage difference in E-field is calculated to be 1.68, 8.23, and 33.9 %. This result would indicate that at high frequency, the EM waves might prefer to

**Fig. 8** Maximum E-field strength through salt water via simulation and laboratory-scale experiment

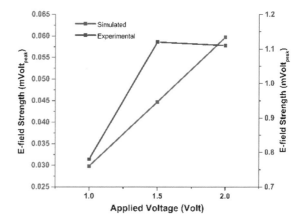

**Fig. 9** Comparative analysis of maximum EM propagation through SS chamber under salt water (model C) at an applied voltage of 1, 1.5, and 2 V

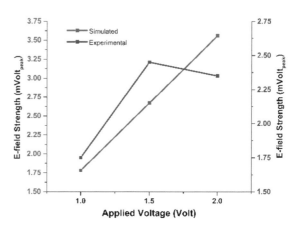

propagate via the SS lining rather than the salt water. It is also noted that the magnitude of electric field decreases slightly by the increase in applied voltage from 1.5 to 2 V. This is due to the unmatched impedance of RF generator to the solenoid in salt water, where the signal loss increases as the applied voltage increased. The percentage deviation between the $E_x$ response clearly shows that the solenoid as a transmitter performs within the agreement to simulation results, especially in the presence of salt water.

## 4 Conclusion

The paper presents the comparative analysis of simulated and experimental results of a solenoid model to be used in conjunction with viscometer. The comparison is made in three different media including air, tap water, and salt water. The $E_x$

response clearly shows that the magnitude of electric field decreases significantly with the presence of stainless steel sample holder, except in the case of salt water due to rapid signal attenuation by salt water conductivity. Further, it is noted that the solenoid used as a transmitter has been best to propagate at an optimum applied voltage of 1.5 V with a designed frequency of 167 MHz in water. This suggests that the main factors enabling the increment in propagation were the transmitter and the operating frequency. Hence, it is concluded that an RF signal generated via a solenoid at frequency of 167 MHz can be transmitted through a SS304 chamber, specifically in seawater. Thus, the observation from these tests can be outlined to further imply in the measurement of viscosity under EM waves.

# References

1. Sheng P, Wen W (2012) Electrorheological fluids: mechanisms, dynamics, and microfluidics application. Annu Rev Fluid Mec 44(1):143–174
2. Winslow WM (1949) Induced fibration of suspensions. J Appl Phys 20:1137–1140
3. Choi HJ, Cho MS, Kim JW, Kim CA, John MS (2001) A yield stress scaling function for electrorheological fluids. Appl Phys Lett 78:3806–3808
4. Wang L, Zhang M, Li J, Gong X, Wen W (2010) Logic control of microfluidics with smart colloid. Lab Chip 10:2869–2874
5. Haroun M, Hassan SA, Ansari A, Kindy NA, Sayed NA, Ali B, Sarma H (2012) Smart nano-EOR process for Abu Dhabi carbonate reservoirs. In: Abu Dhabi International petroleum conference and exhibition, Abu Dhabi, UAE, 11–14 Nov 2012
6. Cui M, Emrick T (2013) Stabilizing liquid drops in nonequilibrium shapes by the interfacial jamming of nanoparticles. Science 25(342):460–463
7. Parasnis DS (1996) Principles of applied geophysics. Springer, Netherlands
8. Abdou AA, Shaw A, Mason A, Al-Shamma'a A, Cullen J, Wylie S, Diallo M (2013) A matched bow-tie antenna at 433 MHz for use in underwater wireless sensor networks. J Phys Conf Ser 450:012048
9. Ghasemi A, Abedi A, Ghasemi F (2012) Propagation engineering in wireless communications. Springer, New York

# MUSIC Algorithm for Imaging of Inhomogeneities Surrounded by Random Scatterers: Numerical Study

**Won-Kwang Park**

**Abstract** We consider an inverse scattering problem in which small dielectric inhomogeneities in two-dimensional space are surrounded by randomly distributed scatterers. We approach this problem using a Multiple Signal Classification (MUSIC) algorithm. This is motivated by the fact that collected Multi-static Response (MSR) matrix data can be represented as an asymptotic expansion formula in the presence of such inhomogeneities. The results obtained by numerical simulations show that MUSIC performs satisfactorily, even under conditions where a significant number of random scatterers affect the data.

## 1 Introduction

One of the main purposes of the inverse scattering problem is to find the locations of unknown inhomogeneities from scattered field data. This problem is challenging, owing to the its ill-posedness. However, it is an interesting problem as it arises in multiple fields, such as physics, medical science, and material engineering. Various detection algorithms have been suggested for approaching this problem, most being based on Newton-type iteration schemes. Related research can be found in [3, 13, 18, 24–27, 31, 37, 39, 42] and references therein. However, in order for such algorithms to be successfully applied, a good initial guess is required, which is close enough to the unknown object. Without this, one might suffer from large computational costs, with the risk of non-convergence issues. Moreover, these schemes require suitable regularization terms, which are highly dependent on the problem at hand; a priori information about unknown inhomogeneities; and the complex calculations of so-called Fréchet derivatives at each iteration step. Even if the above conditions are fulfilled, iteration schemes are very difficult to extend to multiple inhomogeneities.

W.-K. Park (✉)
Department of Mathematics, Kookmin University, 77 Jeongneung-ro,
Seongbuk-gu, Seoul 02707, South Korea
e-mail: parkwk@kookmin.ac.kr

© Springer International Publishing Switzerland 2016
P.J. Soh et al. (eds.), *Theory and Applications of Applied Electromagnetics*,
Lecture Notes in Electrical Engineering 379, DOI 10.1007/978-3-319-30117-4_5

Motivated by this difficulties, alternative noniterative algorithms have been successfully developed and applied to the various inverse problems, such as the MUltiple SIgnal Classification (MUSIC) [2, 4, 5, 7, 8, 15, 21, 22, 33, 35, 36, 43], linear sampling method [12, 14, 16, 17, 23], topological derivative [1, 9, 11, 13, 29, 30], and subspace migration algorithms [20, 28, 32, 34, 38]. Among these, MUSIC (MUltiple SIgnal Classification)-type algorithms have been successfully applied to the imaging of various types of inhomogeneities at fixed single frequency. However, their feasibility has only been confirmed for cases where the background medium is homogeneous. Therefore, the examination of the imaging performance of MUSIC when unknown inhomogeneities are surrounded by random scatterers presents an interesting research subject.

In this paper, we apply a MUSIC-type imaging algorithm to the detection of the locations of small dielectric inhomogeneities that are surrounded by dielectric random scatterers. This is based on the fact that the elements of the so-called Multi-Static Response (MSR) matrix can be represented by an asymptotic expansion formula, owing to the existence of inhomogeneities. For more details, we refer the reader to refer [4]. Using this property, we introduce a MUSIC-type imaging algorithm and perform various numerical simulations.

The rest of this paper is organized as follows. In Sect. 2, we briefly discuss the two-dimensional direct scattering problem and present an asymptotic expansion formula in the presence of small inhomogeneities. In Sect. 3, a MUSIC-type imaging functional is introduced. In Sect. 4, we present various results of numerical simulations, illustrating the effectiveness and limitations of MUSIC. A short conclusion follows in Sect. 5.

## 2  Direct Scattering Problem and Asymptotic Expansion Formula

In this section, we review a two-dimensional direct scattering problem and introduce an asymptotic expansion formula. For a more detailed description, we refer the reader to refer [4]. Let $\Sigma_m$, $m = 1, 2, \ldots, M$ be a dielectric inhomogeneity with a small diameter $r_m$, in the two-dimensional space $\mathbb{R}^2$. Throughout this paper, we assume that every $\Sigma_m$ can be expressed as

$$\Sigma_m = \mathbf{z}_m + r_m \mathbf{B}_m,$$

where $\mathbf{z}_m$ denotes the location of $\Sigma_m$ and $\mathbf{B}_m$ is a simply connected smooth domain containing the origin. For the sake, we let $\Sigma$ be the collection of all $\Sigma_m$. Throughout this paper, we assume that inhomogeneities are well separated from each other, i.e., that there exists $d \in \mathbb{R}$ such that

$$0 < r \ll d \le |\mathbf{z}_m - \mathbf{z}_{m'}|$$

for all $m, m' = 1, 2, \ldots, M$ and $m \ne m'$.

Let us denote $\varDelta_s$, $s = 1, 2, \ldots, S$, be the random scatterer with small a radius $r_s < r$, and let $\varDelta$ be the collection of all $\varDelta_s$. Similarly to the above, we assume that $\varDelta_s$ is of the form

$$\varDelta_s = \mathbf{y}_s + r_s \mathbf{B}_s.$$

and $\varDelta_s \cup \varDelta_{s'} = \emptyset$ for all $s, s' = 1, 2, \ldots, S$ and $s \ne s'$.

In this paper, we assume that all inhomogeneities are characterized by their dielectric permittivity at a given positive angular frequency $\omega = 2\pi/\lambda$, where $\lambda$ denotes the wavelength. Let $\varepsilon_m$, $\varepsilon_s$, and $\varepsilon_0$ be the electric permittivities of $\Sigma_m$, $\varDelta_s$, and $\mathbb{R}^2$, respectively. Then, we can introduce the piecewise-constant electric permittivity $\varepsilon(\mathbf{x})$, such that

$$\varepsilon(\mathbf{x}) = \begin{cases} \varepsilon_m & \text{for} \quad \mathbf{x} \in \Sigma_m, \\ \varepsilon_s & \text{for} \quad \mathbf{x} \in \varDelta_s, \\ \varepsilon_0 & \text{for} \quad \mathbf{x} \in \mathbb{R}^2 \backslash (\bar{\Sigma} \cup \bar{\varDelta}.) \end{cases}$$

For the sake of simplicity, we let $\varepsilon_0 = 1$ and $\varepsilon_m > \varepsilon_s$, for all $m$ and $s$. Hence, we can set the wave number $k = \omega \sqrt{\varepsilon_0} = \omega$.

For a given fixed frequency $\omega$, let

$$u_{\text{inc}}(\mathbf{x}, \boldsymbol{\theta}) = e^{i\omega\boldsymbol{\theta}\cdot\mathbf{x}}$$

be the plane-wave incident field with the incident direction $\boldsymbol{\theta} \in \mathbb{S}^1$, where $\mathbb{S}^1$ denotes a two-dimensional unit circle. Let $u(\mathbf{x}, \boldsymbol{\theta})$ denotes the time-harmonic total field that satisfies the Helmholtz equation

$$\triangle u(\mathbf{x}, \boldsymbol{\theta}) + \omega^2 \varepsilon(\mathbf{x}) u(\mathbf{x}, \boldsymbol{\theta}) = 0,$$

with transmission conditions on the boundaries of $\Sigma_m$ and $\varDelta_s$. It is well known that $u(\mathbf{x}, \boldsymbol{\theta})$ can be decomposed as

$$u(\mathbf{x}, \boldsymbol{\theta}) = u_{\text{inc}}(\mathbf{x}, \boldsymbol{\theta}) + u_{\text{scat}}(\mathbf{x}, \boldsymbol{\theta}),$$

where $u_{\text{scat}}(\mathbf{x}, \boldsymbol{\theta})$ denotes the unknown scattered field that satisfies the Sommerfeld radiation condition

$$\lim_{|\mathbf{x}| \to 0} \sqrt{|\mathbf{x}|} \left( \frac{\partial u_{\text{scat}}(\mathbf{x}, \boldsymbol{\theta})}{\partial |\mathbf{x}|} - i\omega u_{\text{scat}}(\mathbf{x}, \boldsymbol{\theta}) \right) = 0$$

uniformly in all directions $\vartheta = \frac{\mathbf{x}}{|\mathbf{x}|} \in \mathbb{S}^1$. The far-field pattern $u_\infty(\vartheta, \theta)$ of the scattered field $u_{\mathrm{scat}}(\mathbf{x}, \theta)$ is defined on $\mathbb{S}^1$. It can be expressed as

$$u_{\mathrm{scat}}(\mathbf{x}, \theta) = \frac{e^{i\omega|\mathbf{x}|}}{\sqrt{|\mathbf{x}|}} u_\infty(\vartheta, \theta) + o\left(\frac{1}{\sqrt{|\mathbf{x}|}}\right), \quad |\mathbf{x}| \to +\infty.$$

Then, by virtue of result in [10], the far-field pattern $u_\infty(\vartheta, \theta)$ can be written using the following asymptotic expansion formula, which plays a key role in the MUSIC-type algorithm that will be designed in the next section:

$$
u_\infty(\vartheta, \theta) = \frac{\omega^2(1+i)}{4\sqrt{\omega\pi}} \left( \sum_{m=1}^{M} r_m^2(\varepsilon_m - \varepsilon_0)|\mathbf{B}_m| e^{i\omega(\theta-\vartheta)\cdot\mathbf{z}_m} \right.
$$
$$
\left. + \sum_{s=1}^{S} r_s^2(\varepsilon_s - \varepsilon_0)|\mathbf{B}_s| e^{i\omega(\theta-\vartheta)\cdot\mathbf{y}_s} \right). \tag{1}
$$

## 3   MUSIC-type Imaging Algorithm

In this section, we introduce a MUSIC-type algorithm for detecting the locations of small inhomogeneities. For the sake of simplicity, we exclude the constant term $\frac{\omega^2(1+i)}{4\sqrt{\omega\pi}}$ from (1). To proceed, let us consider the eigenvalue structure of the MSR matrix

$$
\mathbb{K} = \begin{bmatrix}
u_\infty(\vartheta_1, \theta_1) & u_\infty(\vartheta_1, \theta_2) & \cdots & u_\infty(\vartheta_1, \theta_N) \\
u_\infty(\vartheta_2, \theta_1) & u_\infty(\vartheta_2, \theta_2) & \cdots & u_\infty(\vartheta_2, \theta_N) \\
\vdots & \vdots & \ddots & \vdots \\
u_\infty(\vartheta_N, \theta_1) & u_\infty(\vartheta_N, \theta_2) & \cdots & u_\infty(\vartheta_N, \theta_N)
\end{bmatrix}.
$$

Suppose that $\vartheta_j = -\theta_j$ for all $j$. Then, $\mathbb{K}$ is a complex symmetric matrix, but is not Hermitian. Therefore, instead of Eigenvalue decomposition, we perform singular value decomposition (SVD) on $\mathbb{K}$ (see [16]):

$$\mathbb{K} \approx \sum_{m=1}^{M} \sigma_m \mathbf{U}_m \mathbf{V}_m^* + \sum_{s=M+1}^{M+S} \sigma_s \mathbf{U}_s \mathbf{V}_s^*, \tag{2}$$

where the superscript $*$ is used to denote the Hermitian. Then, $\{\mathbf{U}_1, \mathbf{U}_2, \ldots, \mathbf{U}_{M+S}\}$ is an orthogonal basis for the signal space of $\mathbb{K}$. Therefore, one can define the projection operator onto the null (or noise) subspace, $\mathbf{P}_{\text{noise}} : \mathbb{C}^{N \times 1} \to \mathbb{C}^{N \times 1}$. This projection is given explicitly by

$$\mathbf{P}_{\text{noise}} := \mathbb{I}_N - \sum_{m=1}^{M+S} \mathbf{U}_m \mathbf{U}_m^*, \tag{3}$$

where $\mathbb{I}_N$ denotes the $N \times N$ identity matrix. For any point $\mathbf{x} \in \mathbb{R}^2$, we define a test vector $\mathbf{f}(\mathbf{x}) \in \mathbb{C}^{N \times 1}$ as

$$\mathbf{f}(\mathbf{x}) = \frac{1}{N} [e^{i\omega\theta_1 \cdot \mathbf{x}}, e^{i\omega\theta_2 \cdot \mathbf{x}}, \ldots, e^{i\omega\theta_N \cdot \mathbf{x}}]^T.$$

Then, by virtue of [4], there exists an $N_0 \in \mathbb{N}$ such that for any $N \geq N_0$, the following statement holds:

$$\mathbf{f}(\mathbf{x}) \in \text{Range}(\mathbb{K}\bar{\mathbb{K}}) \quad \text{if and only if } \mathbf{x} \in \Sigma_m \text{ or } \mathbf{x} \in \Delta_s,$$

for $m = 1, 2, \ldots, M$ and $s = 1, 2, \ldots, S$. This means that if $\mathbf{x} \in \Sigma_m$ or $\mathbf{x} \in \Delta_s$, then

$$|\mathbf{P}_{\text{noise}}(\mathbf{f}(\mathbf{x}))| = 0.$$

Thus, the locations of $\Sigma_m$ and $\Delta_s$ follow from computing the MUSIC-type imaging function

$$\mathcal{I}(\mathbf{x}) = \frac{1}{|\mathbf{P}_{\text{noise}}(\mathbf{f}(\mathbf{x}))|}. \tag{4}$$

The resulting plot of $\mathcal{I}(\mathbf{x})$ will have peaks of large magnitudes at $\mathbf{z}_m \in \Sigma_m$ and $\mathbf{y}_s \in \Delta_s$.

*Remark 3.1* If the size or permittivity of $\Delta_s$ is sufficiently small such that either $r_s \ll r_m$ or $\varepsilon_s \ll \varepsilon_m$, for all $m = 1, 2, \ldots, M$ and $s = 1, 2, \ldots, S$, then the values of the singular value $\sigma_s$ can be negligible. In this case, the locations of the $\Delta_s$ cannot be detected via the map of $\mathcal{I}(\mathbf{x})$. If neither the size nor the permittivity of $\Delta_s$ is that small, then $\sigma_s$ cannot be negligible, i.e., the locations of $\Delta_s$ will be detected via the map of $\mathcal{I}(\mathbf{x})$.

## 4 Results of Numerical Simulations

In this section, the results of some numerical simulations are exhibited in order to examine the imaging performance of MUSIC. The radii of all $\Sigma_m$ and $\Delta_s$ are set to 0.1 and 0.03 (or 0.05), respectively, and permittivities of $\Delta_s$ are selected as random values between 1 and 2. For the incident (and observation) directions, we set

$$\boldsymbol{\theta}_j = -\left[\cos\frac{2\pi(j-1)}{N}, \frac{2\pi(j-1)}{N}\right]^T.$$

The corresponding test configuration is shown in Table 1, and distribution of three inhomogeneities and random scatterers is presented in Fig. 1.

It is worth emphasizing that the dataset of the MSR matrix $\mathbb{K}$ is generated by means of the Foldy-Lax framework in order to avoid *inverse crime*. For more details, we refer the reader to refer [38, 41]. After the generation, the singular value decomposition of $\mathbb{K}$ is performed via the MATLAB command 'svd.' In order to distinguish nonzero singular values of $\mathbb{K}$, a 0.1-threshold scheme is applied, i.e., choosing the first $j$ singular values $\sigma_j$ such that $\frac{\sigma_j}{\sigma_1} \geq 0.1$. For a more detailed description, we refer the reader to refer [35, 36].

Figure 2 shows the distribution of the normalized singular values of $\mathbb{K}$ and map of $\mathcal{I}(\mathbf{x})$ under setting 1. Because the radius and the permittivities of the random scatterers are small, the three nonzero singular values are successfully determined, so one can detect the locations of the $\Sigma_m$ exactly.

Figure 3 shows the distribution of the normalized singular values of $\mathbb{K}$ and map of $\mathcal{I}(\mathbf{x})$ under setting 2. Although some artifacts are included in the map of $\mathcal{I}(\mathbf{x})$, the locations of the $\Sigma_m$ are successfully identified.

**Table 1** Test configuration

| Settings | Value of $N$ | Value of $\lambda$ | Value of $\varepsilon_m$ | Value of $r_s$ |
|---|---|---|---|---|
| Setting 1 | 32 | 0.5 | 5 | 0.03 |
| Setting 2 | 48 | 0.2 | 5 | 0.03 |
| Setting 3 | 48 | 0.4 | 3 | 0.05 |
| Setting 4 | 128 | 0.2 | 3 | 0.05 |

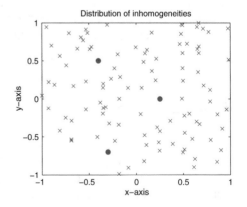

**Fig. 1** Distribution of inhomogeneities (*red-colored circle*) and random scatterers (*blue-colored '×' mark*)

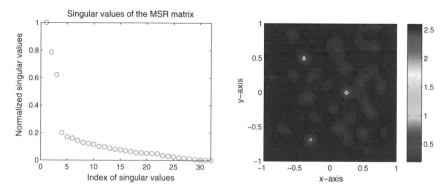

**Fig. 2** Distribution of normalized singular values (*left*) and map of $\mathcal{I}(\mathbf{x})$ (*right*) under setting 1 configuration

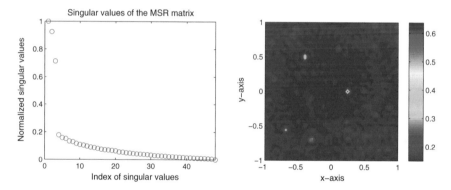

**Fig. 3** Similar to Fig. 2, except under setting 2 configuration

Note that if the values $|\varepsilon_m - \varepsilon_s|$ and $|r_m - r_s|$ are small, then the $\sigma_s$ is no longer negligible. This means that the locations of the $\Lambda_s$ will be identified in the map of $\mathcal{I}(\mathbf{x})$, and this will disturb the identification of the $\Sigma_m$. This is evident in Fig. 4.

For the final example, let us consider the results in Fig. 5. Although $N$ is sufficiently large and the value of $\lambda$ is sufficiently small, it is very difficult to identify the true locations of the $\Sigma_m$.

Based on Figs. 2, 3, 4, and 5, we can conclude that MUSIC is an effective detection technique when the values of the permittivity and the radii of random scatterers are small compared to those of the inhomogeneities. On the other hand, as in the case of our settings 3 and 4, a method of improvement is highly necessary.

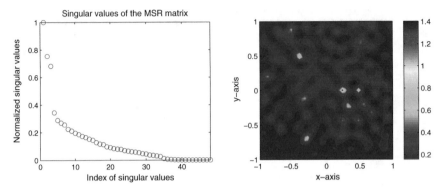

**Fig. 4** Similar to Fig. 2, except under setting 3 configuration

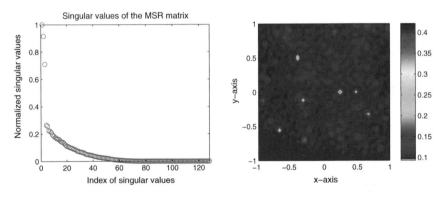

**Fig. 5** Similar to Fig. 2, except under setting 4 configuration

## 5  Conclusion

In this paper, we have considered a MUSIC-type imaging algorithm that is based on an asymptotic expansion formula in the presence of small inhomogeneities and random scatterers. Furthermore, the imaging performance of this MUSIC-type algorithm has been considered when small inhomogeneities are surrounded by random scatterers. Through numerical results, we have observed that MUSIC is an effective technique for identifying inhomogeneities when the random scatterers can be considered negligible, but still required an improvement when the random scatterers cannot be neglected.

In this paper, we focused on the numerical study of MUSIC. A mathematical analysis of the MUSIC-type imaging functional will be considered in future research, by establishing a relationship with Bessel functions of integer order of first kind. We expect that this analysis will illuminate some theoretical properties of MUSIC and suggest some methods of improvements.

Presently, we have considered a purely dielectric case. However, it could be extended to the contrasting purely magnetic case. Moreover, based on the mathematical treatment of the asymptotic formula, the imaging algorithm could be extended to a three-dimensional problem. For more details, we refer the reader to refer [6, 19, 40].

**Acknowledgments** This research was supported by the Basic Science Research Program through the National Research Foundation of Korea (NRF), funded by the Ministry of Education (No. NRF-2014R1A1A2055225)

# References

1. Ahn CY, Jeon K, Ma YK, Park WK (2014) A study on the topological derivative-based imaging of thin electromagnetic inhomogeneities in limited-aperture problems. Inverse Prob 30:105004
2. Ahn CY, Jeon K, Park WK (2015) Analysis of MUSIC-type imaging functional for single, thin electromagnetic inhomogeneity in limited-view inverse scattering problem. J Comput Phys 291:198–217
3. Àlvarez D, Dorn O, Irishina N, Moscoso M (2009) Crack reconstruction using a level-set strategy. J Comput Phys 228:5710–5721
4. Ammari H, Kang H (2004) Reconstruction of small Inhomogeneities from boundary measurements. Lecture notes in mathematics, vol 1846. Springer, Berlin
5. Ammari H, Iakovleva E, Lesselier D (2005) A MUSIC algorithm for locating small inclusions buried in a half-space from the scattering amplitude at a fixed frequency. Multiscale Model Simul 3:597–628
6. Ammari H, Iakovleva E, Lesselier D, Perrusson G (2007) MUSIC type electromagnetic imaging of a collection of small three-dimensional inclusions. SIAM J Sci Comput 29: 674–709
7. Ammari H, Kang H, Lee H, Park WK (2010) Asymptotic imaging of perfectly conducting cracks. SIAM J Sci Comput 32:894–922
8. Ammari H, Garnier J, Kang H, Park WK, Sølna K (2011) Imaging schemes for perfectly conducting cracks. SIAM J Appl Math 71:68–91
9. Ammari H, Garnier J, Jugnon V, Kang H (2012) Stability and resolution analysis for a topological derivative based imaging functional. SIAM J Control Optim 50:48–76
10. Beretta E, Francini E (2003) Asymptotic formulas for perturbations of the electromagnetic fields in the presence of thin imperfections. Contemp Math 333:49–63
11. Burger M, Hackl B, Ring W (2004) Incorporating topological derivatives into level-set methods. J Comput Phys 194:344–362
12. Cakoni F, Colton D (2003) The linear sampling method for cracks. Inverse Prob 19:279–295
13. Carpio A, Rapun ML (2008) Solving inhomogeneous inverse problems by topological derivative methods. Inverse Prob 24:045014
14. Charalambopoulos A, Gintides D, Kiriaki K (2002) The linear sampling method for the transmission problem in three-dimensional linear elasticity. Inverse Prob 18:547–558
15. Chen X, Zhong Y (2009) MUSIC electromagnetic imaging with enhanced resolution for small inclusions. Inverse Prob 25:015008
16. Cheney M (2001) The linear sampling method and the MUSIC algorithm. Inverse Prob 17:591–595
17. Colton D, Haddar H, Monk P (2002) The linear sampling method for solving the electromagnetic inverse scattering problem. SIAM J Sci Comput 24:719–731

18. Dorn O, Lesselier D (2006) Level set methods for inverse scattering. Inverse Prob 22: R67–R131
19. Gdoura S, Lesselier D, Chaumet PC, Perrusson G (2009) Imaging of a small dielectric sphere buried in a half space. ESAIM Proc 26:123–134
20. Hou S, Huang K, Sølna K, Zhao H (2009) A phase and space coherent direct imaging method. J Acoust Soc Am 125:227–238
21. Joh YD, Park WK (2013) Structural behavior of the MUSIC-type algorithm for imaging perfectly conducting cracks. Prog Electromagn Res 138:211–226
22. Joh YD, Kwon YM, Park WK (2014) MUSIC-type imaging of perfectly conducting cracks in limited-view inverse scattering problems. Appl Math Comput 240:273–280
23. Kirsch A, Ritter S (2000) A linear sampling method for inverse scattering from an open arc. Inverse Prob 16:89–105
24. Kress R (1995) Inverse scattering from an open arc. Math Methods Appl Sci 18:267–293
25. Kress R, Serranho P (2005) A hybrid method for two-dimensional crack reconstruction. Inverse Prob 21:773–784
26. Kress R, Serranho P (2007) A hybrid method for sound-hard obstacle reconstruction. J Comput Appl Math 204:418–427
27. Mönch L (1996) On the numerical solution of the direct scattering problem for an open sound-hard arc. J Comput Appl Math 17:343–356
28. Park WK (2010) On the imaging of thin dielectric inclusions buried within a half-space. Inverse Prob 26:074008
29. Park WK (2012) Topological derivative strategy for one-step iteration imaging of arbitrary shaped thin, curve-like electromagnetic inclusions. J Comput Phys 231:1426–1439
30. Park WK (2013) Multi-frequency topological derivative for approximate shape acquisition of curve-like thin electromagnetic inhomogeneities. J Math Anal Appl 404:501–518
31. Park WK (2013) Shape reconstruction of thin electromagnetic inclusions via boundary measurements: level-set method combined with topological derivative. Math Probl Eng 2013:125909
32. Park WK (2014) Analysis of a multi-frequency electromagnetic imaging functional for thin, crack-like electromagnetic inclusions. Appl Numer Math 77:31–42
33. Park WK (2015) Asymptotic properties of MUSIC-type imaging in two-dimensional inverse scattering from thin electromagnetic inclusions. SIAM J Appl Math 75:209–228
34. Park WK (2015) Multi-frequency subspace migration for imaging of perfectly conducting, arc-like cracks in full- and limited-view inverse scattering problems. J Comput Phys 283:52–80
35. Park WK, Lesselier D (2009) Electromagnetic MUSIC-type imaging of perfectly conducting, arc-like cracks at single frequency. J Comput Phys 228:8093–8111
36. Park WK, Lesselier D (2009) MUSIC-type imaging of a thin penetrable inclusion from its far-field multi-static response matrix. Inverse Prob 25:075002
37. Park WK, Lesselier D (2009) Reconstruction of thin electromagnetic inclusions by a level set method. Inverse Prob 25:085010
38. Park WK, Lesselier D (2012) Fast electromagnetic imaging of thin inclusions in half-space affected by random scatterers. Waves Random Complex Media 22:3–23
39. Santosa F(1996) A level-set approach for inverse problems involving obstacles. ESAIM Control Optim Calc Var 1:17–33
40. Song R, Chen R, Chen X (2012) Imaging three-dimensional anisotropic scatterers in multi-layered medium by MUSIC method with enhanced resolution. J Opt Soc Am A 29:1900–1905
41. Tsang L, Kong JA, Ding KH, Ao CO (2001) Scattering of Electromagnetic Waves: Numerical Simulations. John Wiley, New York
42. Ventura G, Xu JX, Belytschko T (2002) A vector level set method and new discontinuity approximations for crack growth by EFG. Int J Numer Methods Eng 54:923–944
43. Zhong Y, Chen X (2007) MUSIC imaging and electromagnetic inverse scattering of multiple-scattering small anisotropic spheres. IEEE Trans Antennas Propag 55:3542–3549

# Frequency Reconfigurable Aperture-Coupled Microstrip Array Antenna Using Periodic Defected Ground Structures

**M.A. Aris, M.T. Ali and N.H. Abd Rahman**

**Abstract** In this paper, a design of $2 \times 2$ frequency reconfigurable microstrip array antenna using Defected Ground Structures (DGS) with aperture-coupled feed line for outdoor mobile applications. Two periodical dumbbell geometries with different dimensions etched on the ground plane. Each dimension of DGS structures etched on the ground plane resonant at two different frequency bands, (7.28–7.73 GHz) and (8.55–9.12 GHz), respectively. The reconfigurability of the antenna is based on the transition modes of switch on the feed line. When the switch is in "OFF" mode, the antenna resonant is at 7.5 GHz and covers 7.28–7.73 GHz frequency band while the antenna resonant is at 8.85 GHz in "ON" mode and covers 8.55–9.13 GHz frequency band. Switching modes for the preliminary prototype are presented in open and short condition of the feed line. The reconfigurable array antenna simulated using Computer Simulation Technology (CST). Results between simulation and measurement provide good agreement of the designed antenna. This new approach offers simple design and small volume of reconfigurable microstrip array antenna for future outdoor mobile applications.

**Keywords** Defected Ground Structures (DGS) · Reconfigurable antenna · Array · Aperture-coupled

M.A. Aris (✉) · M.T. Ali · N.H. Abd Rahman
Antenna Research Group (ARG), Microwave Technology Center (MTC),
Faculty of Electrical Engineering, University Teknologi MARA Shah Alam,
Shah Alam, Selangor 40450, Malaysia
e-mail: Mohda474@tganu.uitm.edu.my

M.T. Ali
e-mail: mizi732002@yahoo.com

N.H. Abd Rahman
e-mail: nurulhuda0340@salam.uitm.edu.my

M.A. Aris
Universiti Teknologi MARA Terengganu, Dungun, Terengganu 23000, Malaysia

© Springer International Publishing Switzerland 2016
P.J. Soh et al. (eds.), *Theory and Applications of Applied Electromagnetics*,
Lecture Notes in Electrical Engineering 379, DOI 10.1007/978-3-319-30117-4_6

61

# 1  Introduction

Based on its versatility, frequency reconfigurable microstrip antenna has received high attention in order to support wide areas of wireless communication systems. This high demand is supported by numerous reasons, such as able to support more than one wireless standard, minimize cost, simple integration, and good isolation between different wireless frequencies [1]. However, due to low gain, very narrow bandwidth of microstrip circuit design reconfigurable antenna in single element is not significant for outdoor wireless systems. Therefore, in order to generate high directivity better bandwidth, frequency reconfigurable antenna with array structure has been developed by many researchers and finely designed to support multifrequency bands wireless communication systems, for example, Wi-Fi, WLAN, WiMAX, and lately for LTE [2, 3]. Unfortunately, frequency reconfigurable antenna array has limited only for indoor wireless network applications, due to its very complex switching and biasing network [4]. In addition, limited array elements and very complex designs such as back-to-back structure and stacked design [5, 6] also contribute main restriction to design antenna array for long distance wireless communication.

Therefore, this paper proposes a $2 \times 2$ microstrip frequency reconfigurable antenna array using periodic Defected Ground Structure (DGS) with aperture-coupled feed line method for outdoor mobile applications. The antenna was designed and fabricated on the substrate Rogers RO3006 with permittivity of 6.15 and thickness of 0.64 mm. The aperture-coupled feed line is selected due to its good background, such as it offers independent optimization of the feed line and the radiating element, low interference between feed line and patch element, and better bandwidth [6]. Defected Ground Structure (DGS) method has potential to offer innovative approach in designing frequency reconfigurable microstrip array antenna based on two main characteristics of DGS, slow wave propagation in Pass band and Band Stop [7], where these criteria are able to reject unwanted frequency band and minimize coupling effect for antenna array [8].

In this research, two identical dumbbell Defected Ground Structure shapes have been chosen and optimized to resonate at 7.5 and 8.85 GHz. The antenna was simulated using Computer Simulation Technology (CST). In this design, the switching concept for reconfigurability refers to open and short condition of antenna feed line. Finally, prototype reconfigurable antenna was fabricated and measured to validate its performance.

# 2  Antenna Design

## 2.1  Single Element Microstrip Antenna

At the beginning of the experimental analysis, systematical steps have been conducted to achieve the main objective. First step is to design a single element antenna

**Fig. 1** Diagram of single
element for frequency
reconfigurable microstrip
patch antenna with
aperture-coupled feed line and
periodic dumbbell DGS
structure

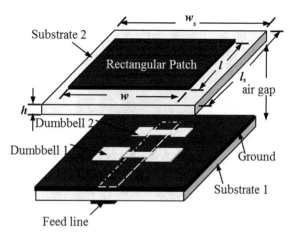

then followed by 2 × 2 array antenna. The initial dimension of the single element
rectangular patch antenna length (l) and width (w) is calculated using Eqs. (1)–(4)
by considering resonant frequency of 8.2125 GHz. Then, two periodic dumbbell
shapes with different sizes etched on the ground layer, exactly located under the
single rectangular radiator to control the desired frequencies. The complete design
for the single element is shown in Fig. 1; where the rectangular radiating element is
placed on top of substrate 2, the ground layer is placed on top of substrate 1. The
size of the periodic dumbbell DGS structures was obtained from optimization
process so that the antenna will resonate at the two desired frequencies. As shown
in Fig. 1, dumbbell 1 (in the middle of the ground layer) resonates at 7.5 GHz,
while dumbbell 2 (at the top of dumbbell 1) resonates at 8.85 GHz. The complete
dumbbell DGS structures on the ground layer are depicted in Fig. 2a, and feed line
with switch concept location is illustrated in Fig. 2b. Meanwhile, the length of the
"$S_w$" is equal to the dimension of proposed PIN diode which will be used as the
switch. All dimensions for the designed single element are listed in Table 1.

$$w = \frac{1}{2f_r\sqrt{\mu_o\varepsilon_o}}\sqrt{\frac{2}{\varepsilon_r + 1}} \tag{1}$$

$$l = \frac{1}{2f_r\sqrt{\varepsilon_{\text{reff}}}\sqrt{\mu_o\varepsilon_o}} - 2\Delta L \tag{2}$$

$$\Delta L = \frac{0.412}{h}\frac{(\varepsilon_{\text{reff}} - 0.3)(\frac{W}{h} + 0.264)}{(\varepsilon_{\text{reff}} - 0.258)(\frac{W}{h} - 0.8)} \tag{3}$$

$$\varepsilon_{\text{reff}} = \frac{\varepsilon_r + 1}{2} + \frac{\varepsilon_r - 1}{2}\left[1 + 12\frac{h}{W}\right]^{-1/2} \tag{4}$$

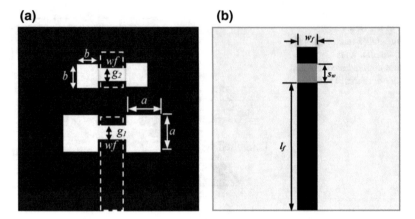

**Fig. 2** **a** Periodic dumbbell DGS structures etched on the ground layer and **b** Feed line and position of switch for frequency reconfigurability

**Table 1** Dimension of designed antenna

| Antenna parameters | Value (mm) |
|---|---|
| l | 9.28 |
| w | 10 |
| $l_s$ | 15 |
| $w_s$ | 15 |
| h | 0.64 |
| wf | 0.995 |
| a | 2.36 |
| b | 0.84 |
| g1 | 0.4 |
| g2 | 0.25 |
| Air gap | 1 |
| $s_w$ | 0.9 |
| lf | 8.32 |

## 2.2  2 × 2 Microstrip Array Antennas

The next step is to design the 2 × 2 microstrip array antenna with the distance of λ/2 between the patch elements and is shown in Fig. 3. A corporate feed network connected to a 50 Ω feed line is used as power divider between antenna elements. The 50 Ω transmission line consists of 70.7 Ω quarter wave transformers with 100 Ω power divider. Meanwhile, four pairs of DGS structures on the ground plane etched with constant dimension as obtained from single element. Dimension of the feed network for 2 × 2 antenna array is depicted in Fig. 4a, b for different resonant frequencies.

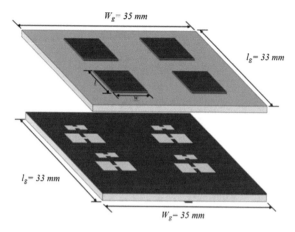

**Fig. 3** Top layer with 2 × 2 rectangular patch antenna and four pairs of DGS structure on the ground plane

**(a)**                **(b)**

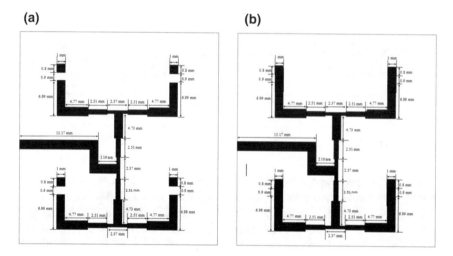

**Fig. 4 a** Feed line with open condition for "OFF" mode and **b** feed line with "ON" mode

## 3 Simulation and Measurement Results

### 3.1 Simulation Result Using Computer Simulation Technology (CST) for Single Element

The proposed antenna is simulated using Computer Simulation Technology (CST). Simulation and optimization started with single element antenna. As shown in Fig. 5a, b, two resonant frequencies were obtained from the designed antenna based

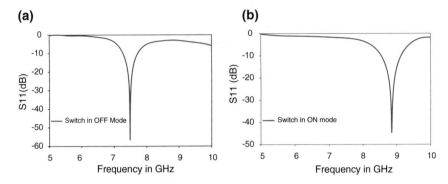

**Fig. 5** **a** Simulated result of S11 in "OFF" mode and **b** simulated result of S11 in "ON" mode

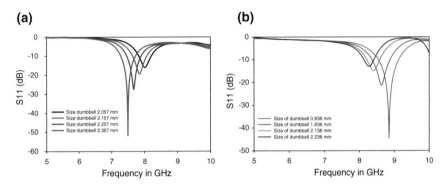

**Fig. 6** **a** Parametric analysis in "OFF" and **b** parametric analysis in "ON" mode

on the switch mode and selected DGS's dimension etched on the ground plane. During OFF mode, the antenna operates at 7.5 GHz with S11–56.58 dB, while in ON mode the antenna operates at 8.85 GHz with S11–44.64 dB. The simulated results for S11 indicate a good match obtained from the proposed antenna. Interestingly, the designed antenna could cover very wideband frequency band at two desired frequencies 450 MHz (7.28–7.73 GHz) and 580 MHz (8.55–9.13 GHz), respectively.

The functionality of DGS structure in producing two desired frequencies is validated through parametric analysis. The analysis is carried out on the gap junction for each DGS shape specified at 0.4 mm with the length of gap follows the width of feed line. The parametric analysis was focused mainly on the effect of varying the square arm of dumbbell structure. Figure 6a shows that decreasing the size of the square arm of the bottom dumbbell will significantly increase the frequency, while Fig. 6b indicates vice versa.

## 3.2 Simulation Result Using Computer Simulation Technology (CST) for 2 × 2 Array Antenna

Similarly 2 × 2 array is simulated using Computer Simulation Technology (CST) and is fabricated with different modes "OFF" and "ON." To confirm the antenna is workable in desired band, simulated and fabricated results are compared and are shown in Fig. 7a in "OFF" mode, while Fig. 7b illustrates the comparison in "ON" mode. During "OFF" mode, the resonant frequency occurs at 7.39 GHz

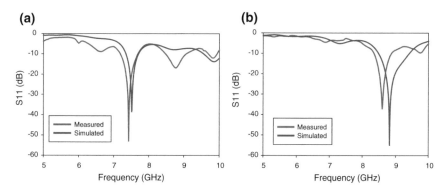

**Fig. 7** **a** Simulated and fabricated results in "OFF" mode and **b** fabricate results in "ON" mode

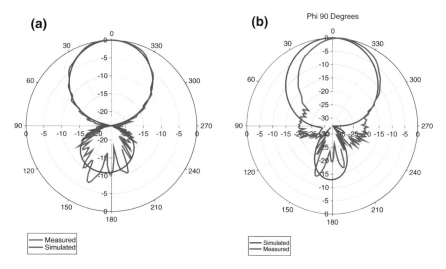

**Fig. 8** **a** Simulated and measured radiation pattern in "OFF" mode and **b** simulated and measured radiation pattern in "ON" mode

with S11–53.20 dB; this result shifts 150 MHz from simulated result. In "ON" mode the fabricated antenna resonate at 8.75 GHz with S11–37.41 dB where shift 100 MHz from simulated result. However, the fabricated antenna still works in expected bandwidth to support meteorology and radiolocation applications. Figure 8a, b shows the comparison results between simulated and measured in normalized form for radiation pattern in "OFF" and "ON" modes, respectively.

The performance of the 2 × 2 antenna array antenna is summarized in Table 2. Figure 9 illustrates the fabricated 2 × 2 antenna array. Referring to Table 2, the efficiency and gain directivity of fabricated antenna slightly decrease compared to simulated results.

**Table 2** Performance of proposed antenna array

| | Switch modes | | | |
|---|---|---|---|---|
| | Simulated | | Measured | |
| | OFF | ON | OFF | ON |
| Center frequency (GHz) | 7.5 | 8.85 | 7.39 | 8.75 |
| Efficiency (%) | 91.88 | 91.92 | 79.62 | 94.62 |
| Directivity (dBi) | 9.62 | 11.1 | 7.92 | 10.49 |
| Bandwidth (MHz) | 446.5 | 573.15 | 439.479 | 492.847 |
| S11 (dB) | −38.77 | −55.23 | −53.20 | −37.41 |

**(a)**                                    **(b)**

**(c)**                                    **(d)**

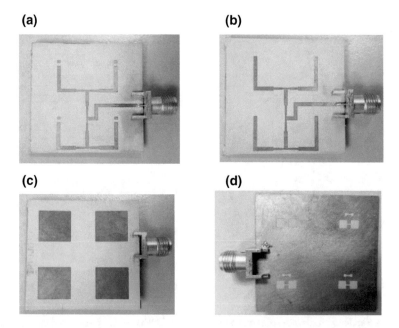

**Fig. 9** **a** Feed line network in "OFF" mode for fabricated antenna, **b** feed line network in "ON" mode for fabricated antenna, **c** rectangular patch fabricated antenna, and **d** ground plane with DGS structures

From simulation and fabrication process, large back lobe from the proposed antenna requires additional technique to reduce it, and we noticed that this lobe size is contributed by DGS structure. Meanwhile, the shifted frequency obtained from fabricated antenna and deterioration of efficiency and directivity possibly comes from fabrication errors.

## 4 Conclusions

From the research, microstrip antenna using DGS structure provides simple structure for microstrip antenna, mainly in designing reconfigurable antenna array, and by varying the size of the DGS shape the desired resonant frequencies are better generated. In addition, the switching concept proposed at the feed line is offering lesser number of reconfiguration switching systems. However, the numbers of switches are significant with the increment of radiator elements.

**Acknowledgments** The authors would like to thank the Faculty of Electrical Engineering (Antenna Research Centre), Universiti Teknologi MARA (UiTM), Malaysia, and Ministry of Higher Education (MOHE), Malaysia, for the financial support under Fundamental Research Grant Scheme (FRGS 600-RMI/FRGS 5/3).

## References

1. Christodoulou CG, Tawk Y, Lane SA, Erwin SR (2012) Reconfigurable antennas for wireless and space applications. Proc IEEE 100(7):2250–2261
2. Qin PY, Jay Guo Y, Ding C (2013) A dual-band polarization reconfigurable antenna for WLAN systems. IEEE Trans Antennas Propag
3. Li Z, Ahmed E, Eltawil AM, Member S, Cetiner BA (2015) A beam-steering reconfigurable antenna for WLAN applications. IEEE Trans Antennas Propag 63(1):24–32
4. Costantine J, Tawk Y, Christodoulou CG, Lyke JC, Member S, De Flaviis F, Besoli AG, Barbin SE (2012) Analyzing the complexity and reliability of switch-frequency-reconfigurable antennas using graph models. IEEE Trans Antennas Propag 60(2):811–820
5. Ramli N, Ali MT, Yusof AL, Ya'Acob N (2013) Frequency reconfigurable stacked patch microstrip antenna (FRSPMA) for LTE and WiMAX applications. In: 2013 international conference computing, management and telecommunications ComManTel 2013, pp 55–59
6. Ali MT, Salleh MKM, Rusli MHM (2014) Reconfigurable gap-coupled back-to-back truncated rhombus-like slotted patch antenna with steerable beams, pp 338–342
7. Arya AK, Kartikeyan MV, Patnaik A (2010) Defected ground structure in the perspective of microstrip antennas: a review. Frequenz
8. Woo DJ, Lee JW, Lee TK (2008) Multi-band rejection DGS with improved slow-wave effect. In: Proceedings of 38th European Microwave Conference EuMC 2008, vol 1, Oct, pp 1342–1345

# Design of a SIW-Based Microstrip Diplexer Using $TM_{010}$ Circular Cavity

Noor Hasimah Baba, Aziati Husna Awang, M.T. Ali, M.A. Aris
and Hizamel Mohd Hizan

**Abstract** This paper describes the application of a Substrate Integrated Waveguide (SIW) for the design of a microstrip diplexer. The diplexer is composed of two SIW circular cavity filters designed individually and combined together using a T-junction with center frequencies at 2.4 and 2.6 GHz, respectively. The working mode for the circular cavity structures is $TM_{010}$ mode. To prove the concept, the diplexer is fabricated using Rogers RT5880 with dielectric constant, $\varepsilon_r = 2.2$ and height of substrate 1.57 mm. The performances of the proposed design are verified through both simulation and measurement.

**Keywords** Circular cavity · Diplexer · T-junction · Substrate Integrated Waveguide · SIW

## 1 Introduction

In last few decades, rapid development in communication systems such as GSM, WLAN, LTE, and satellite applications has introduced many wireless products for millimeter-wave applications. Each of these systems uses passive microwave devices to operate such as filter and diplexer. Generally, both of these devices are larger compared to other components in the system. Therefore, there will be a great interest to make these devices to become smaller and more compact. The diplexer normally composed of two filters is a two-way three-port network commonly used for separation of received and transmits signals served by a common antenna [1]. Traditional structure diplexers suffer from disadvantages such as being bulky,

N.H. Baba (✉) · A.H. Awang · M.T. Ali · M.A. Aris
Faculty of Electrical Engineering, Universiti Teknologi MARA Malaysia,
40450 Shah Alam, Selangor, Malaysia
e-mail: nhasimah@salam.uitm.edu.my

H.M. Hizan
Advanced Physical Technologies Lab, Telekom Research & Development Sdn Bhd,
Cyberjaya, Selangor, Malaysia

© Springer International Publishing Switzerland 2016
P.J. Soh et al. (eds.), *Theory and Applications of Applied Electromagnetics*,
Lecture Notes in Electrical Engineering 379, DOI 10.1007/978-3-319-30117-4_7

costly, and difficult to fabricate. In addition, it cannot be integrated with the mm-wave planar integrated circuits of the transceiver, and the degradation of performance can be substantial for transition from a waveguide to planar integrated circuits [2, 3].

The diplexer and its operation principle are shown in Fig. 1b. In transmitting mode, two signals with different frequencies are injected in port 1: One of them exits from port 2, while the other one from port 3. In receiving mode, ports 2 and 3 receive two different signals, which exit from port 1. Over the past decade, many works have been carried out to replace all-metallic waveguide structures with SIW circuits. Substrate Integrated Waveguide (SIW) is a type of dielectric-filled waveguide which is implemented in a planar substrate with linear arrays of metallic vias. Those vias are used to realize metal edge wall that connects the two ground planes of the substrate. The SIW offers advantages such as high Q factor, low insertion loss, low cost, and high power-handling capacity [4, 5]. Apart from the

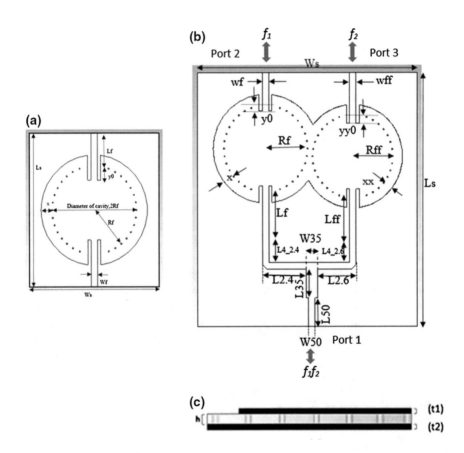

**Fig. 1** **a** Top view of SIW filter **b** top and **c** cross-sectional views of SIW diplexer

well-known advantages, this technique widely appears in various conventional passive microwave components and devices such as filters, couplers, power dividers, and antennas [6–9].

## 2   Methodology

Initial step in designing the diplexer is to first design and simulate the individual filters satisfying the specifications desired. The two filters have been designed at center frequencies of 2.4 and 2.6 GHz, respectively. Secondly, these filters are combined together using suitable matching network. In this case, a microstrip T-junction power divider followed by a 50 Ω microstrip line with inset feed is used to directly excite the filters. Configurations of the proposed filter and diplexer with the transitions are shown in Fig. 1. The topology comprises one metallic top layer (t1), one conductive ground layer (t2), two circular cavity resonators, inset couplings, and linear arrays of via holes to form a circular via wall around the cavity resonators. Measured results are presented and compared to those simulated by 3D Electromagnetic Simulation (CST) software package.

### 2.1   *Network Synthesis of Diplexer*

To begin with, a network that meets Table 1 design specification is synthesized using standard filter theory starting with a low-pass filter prototype having Chebyshev response [10]. The equivalent circuit of the first-order diplexer is shown in Fig. 2. In this network, the admittance inverters J01 and J12 are used as the input and output coupling of the filter.

The shunt connected resonators are calculated using Eqs. (1) and (2), as shown below:

$$L_1' = \frac{1}{\alpha C_1 \omega_0} \tag{1}$$

| Table 1 First-order TM$_{010}$ SIW filter specifications | Filter specifications | |
|---|---|---|
| | Center frequency, $f_0$ | 2.4 and 2.6 GHz |
| | Passband bandwidth (PBW) | 20 MHz |
| | Passband return loss ($L_R$) | $\geq$10 dB |
| | Stopband insertion loss ($L_A$) | >10 dB at $f_0 \pm$ 100 MHz (SBW) |

**Fig. 2** Equivalent circuit for
first-order diplexer

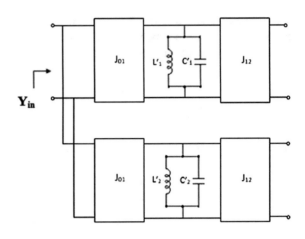

$$C'_1 = \frac{\alpha C_1}{\omega_0} \qquad (2)$$

where $\alpha$ is the bandwidth scaling factor and $\omega_0$ is the midband frequency. All the
element values for the diplexer are shown in Table 2.

The simulated responses of the first-order diplexer having center frequencies at
2.4 and 2.6 GHz with 10 dB passband return loss bandwidth of 20 MHz are shown
in Fig. 3.

**Table 2** Element values for
single-mode diplexer

| Element | Value (2.4 GHz) | Value (2.6 GHz) |
|---|---|---|
| $L'_1 = L'_2$ | 0.8289 pH | 0.7063 pH |
| $C'_1 = C'_2$ | 5.3052 nF | 5.3052 nF |
| $J_{01} = J_{12}$ | 1 | 1 |

**Fig. 3** Equivalent circuit
simulation results

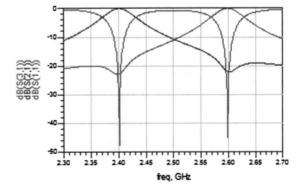

## 2.2  *Electromagnetic Simulation*

The fundamental parameter in designing the cavity of the SIW filter is the resonant frequency. The radius of the circular SIW cavity is approximated according to the resonant frequency using the following formula [10], and then the radius is optimized by electromagnetic software.

$$\left(f_r^{\text{filter}}\right)^{TM^z}_{010} = \frac{c}{2\pi\sqrt{\mu_r \varepsilon_r}} \left(\sqrt{\frac{2.4049}{a_{\text{filter}}}}\right)^2 \tag{3}$$

where c is the speed of light in free space; $\mu_r$ is the relative permeability while $\varepsilon_r$ is the dielectric permeability of the substrate, respectively. The value 2.4049 is the first zero of the Bessel function and $a_{\text{filter}}$ is the radius of the SIW filter. The design rules related to the pitch and via diameters to ensure that the radiation loss is kept at negligible level are already given in [11]. In this design, the matching from 50 $\Omega$ microstrip line to the SIW cavity resonator section is achieved by the inset excitation structure which converts quasi-TEM mode propagating in microstrip line to the TM$_{010}$ mode of the SIW cavity resonator. From the simulation, it is observed that the depth inset coupling parameter y0 of Fig. 1 is considered critical and needs to be carefully designed and optimized. Figure 4 shows the variation of the length y0 of the SIW diplexer, indicating that increasing the depth slot from 1.95 to 3.95 mm will increase the coupling and the return loss values. The optimized geometric parameter of the proposed structure is shown in Table 3.

**Fig. 4** Effect of inset length y0 of SIW diplexer

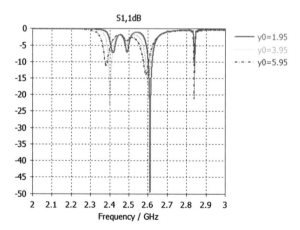

**Table 3** Geometric dimensions of first-order diplexer

| Symbol | 2.4 GHz Value (mm) | 2.6 GHz Value (mm) | Symbol | Value (mm) |
|--------|---------------------|---------------------|--------|------------|
| Rf | 30.33 | 29.12 | $h$ | 1.57 |
| wf | 4.65 | 4.65 | Ls | 177.80 |
| yo | 3.96 | 36.36 | Ws | 160 |
| y0 | 3.95 | 6.17 | $t$ | 0.035 |
| Lf | 22.74 | 20.99 | | |
| $x$ | 7.81 | 7.21 | | |

## 3   Results

To prove the concept, the diplexer structure is fabricated using RT-5880 substrate with dielectric constant, $\varepsilon_r = 2.2$ and thickness of 1.57 mm, and a photograph of the diplexer prototype is shown in Fig. 7. When measuring the S parameters of the diplexer, one port is matched with a standard 50 Ω load, and the other two ports are connected to a vector network analyzer (VNA). Figure 5 shows the simulated and measured S parameter response of 2.4 and 2.6 GHz SIW filters. The simulated and measured results of the SIW diplexer are given in Fig. 6. From this plot, the simulation return losses for the lower and upper channels are greater than 28.68 and 17.16 dB, respectively. The return losses measured at the ports remain greater than 12.24 and 17.86 dB within the associated channels. The measured minimum insertion losses show a slight deviation from 1.42 and 1.07 dB to 2.24 and 3.17 dB at each channel. Measured result shows that the frequencies for the lower and upper channels are shifted for about 30 MHz. The shifting of the frequencies is, however, considered acceptable and is probably due to errors which occurred during the fabrication process and tolerances of the substrate dielectric constant. The difference may also be due to extra loss from the SMA connectors. Consider the reasons of the errors, the presence of air gap may be the possible reason, because higher energy

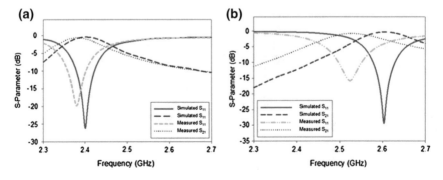

**Fig. 5  a** Simulated and measured S parameters of 2.4 GHz SIW filter **b** simulated and measured S parameters of 2.6 GHz SIW filter

**Fig. 6** Simulated and
measured S parameters
response of SIW diplexer

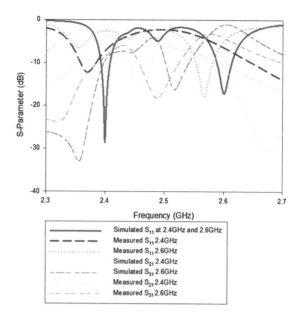

| | |
|---|---|
| ——————— | Simulated S$_{11}$ at 2.4GHz and 2.6GHz |
| — — — — | Measured S$_{11}$ 2.4GHz |
| ·················· | Measured S$_{11}$ 2.6GHz |
| —————— | Simulated S$_{21}$ 2.4GHz |
| — — — — — | Simulated S$_{31}$ 2.6GHz |
| —————— | Measured S$_{21}$ 2.4GHz |
| — —— — —— — | Measured S$_{31}$ 2.6GHz |

**Fig. 7** Photograph of the
fabricated SIW diplexer

losses affect the efficiency of the device. Also the metallic vias used were not
perfect conductors and seems they have different values of dielectric, while the
dielectric for PEC is infinite [12].

Figure 8 plots the measured results of the isolation between ports 2 and 3. The
diplexer provides isolation better than 15 dB at both channels. This moderate performance is caused by the coupling effects that the adjacent resonators introduce.
The main limitation of the SIW technology is the maximum isolation that a dis-
continuous wall made of via holes can provide between adjacent resonators, producing undesirable cross-couplings [13]. However, even with non-perfect isolation
the design of a diplexer is possible, and with a good fabrication process, the integrated structure can give a better performance. In addition, in order to improve the
response and bandwidth of the structure, higher degree of filter can be employed.

**Fig. 8** Measured isolation of
the SIW diplexer

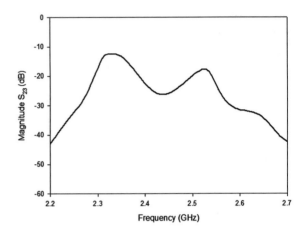

## 4 Conclusion

A new prototype of compact SIW microstrip diplexer is proposed employing single-mode circular cavity resonators having center frequencies at 2.4 and 2.6 GHz, and which is proposed for LTE/WiFi applications. Good agreement between simulation and measurement is achieved. The diplexer features low insertion loss, compact, simple structure, and easy to connect to other circuits. This design provides an alternative solution for the uplink/downlink RF front-end subsystem that is essential for wireless communications systems.

**Acknowledgments** This work was supported by the Research Acculturation Collaborative Effort (RACE) Funds, Ministry of Education (No. 600-RMI/RACE 16/6/2(8/2013).

## References

1. He J, Gao K, Shao Z (2012) A novel compact Ka-Band high-rejection diplexer based on substrate integrated waveguide. In: International conference computational problem-solving (ICCP), pp. 193–197
2. Kordiboroujeni Z, Bornemann J (2013) Mode matching design of substrate integrated waveguide diplexers. In: IEEE MTT-S international microwave symposium digest (IMS)
3. Athanasopoulos N, Makris D, Voudouris K (2011) Development of a 60 GHz substrate integrated waveguide planar diplexer. In: 2011 IEEE MTT-S international microwave workshop series on millimeter wave integration technologies
4. Tang HJ, Hong W, Chen J-X, Luo GQ, Wu K (2007) Development of millimeter-wave planar diplexers based on complementary characters of dual-mode substrate integrated waveguide filters with circular and elliptic cavities. IEEE Trans Microw Theory Tech 55(4)
5. Tang HJ, Hong W, Hao ZC, Chen JX, Wu K (2005) Optimal design of compact millimeter-wave SIW circular cavity filters. Electron Lett 41(19)
6. Dashti H, Shahabadi M, Neshati MH (2013) SIW cavity-backed slot antennas with improved gain. In: 21st Iranian conference electrical engineering (ICEE)

7. Bozzi M (2012) Substrate integrated waveguide (SIW): an emerging technology for wireless systems In: Proceedings of APMC 2012, Kaohsiung, Taiwan, Dec 4–7
8. Garcia-Lamperez A, Salazar-Palma M, Yeung SH (2014) SIW compact diplexer. In: IEEE MTT-S international microwave symposium (IMS)
9. Hong W (2005) Development of microwave antennas, components and subsystems based on SIW technology. In: Proceedings of 2005 IEEE international symposium on microwave, antenna, propagation and EMC technologies, pp 14–17
10. Hunter IC (2001) Theory and design of microwave filters. Institution of Electrical Engineers, London
11. Hizan HM, Hunter IC, Abunjaileh AI (2010) Integrated SIW filter and microstrip antenna. In: Proceedings of the 40th European microwave conference 28–30 September 2010, Paris, France
12. Balanis CA (2005) Antenna theory: analysis and design, 3rd edn. Wiley, Hoboken
13. Nawaz MI, Huiling Z (2014) Substrate integrated waveguide (SIW) to microstrip transition at X-band. In: Proceedings of the 2014 international conference on circuits, systems and control, pp 61–63

# Planar Textile Antennas Performance Under Wearable and Body Centric Measurements

Kamilia Kamardin, Mohamad Kamal A. Rahim,
Noor Asmawati Samsuri, Mohd Ezwan Jalil
and Noor Azurati Ahmad

**Abstract** This study proposes three types of planar textile antennas namely planar straight dipole, diamond dipoles and CPW monopole for body centric communication. All the proposed antennas are made from entirely textile where fleece and Shieldit fabrics are used as substrates and conducting parts, respectively. The proposed antennas have been thoroughly measured under wearable and body centric measurements. Investigations including bending, wetness and SAR were performed to test the performance of the proposed antennas for practical realization in body centric applications. Bending was found not to give any obvious performance deviation. However, wetness has caused major performance disruption since the antennas are made from non-waterproof material. Nevertheless, the original performance is recovered once the antennas are dried out. SAR investigation was also performed, and both antennas give significant SAR values when positioned near to human body.

**Keywords** Textile antenna · Bending · Wetness · Specific absorption rate

K. Kamardin (✉) · N.A. Ahmad
Computer Systems Engineering Group, Advanced Informatics School,
Universiti Teknologi Malaysia, 54100 Kuala Lumpur, Malaysia
e-mail: kamilia@utm.my

N.A. Ahmad
e-mail: azurati@utm.my

M.K.A. Rahim · N.A. Samsuri · M.E. Jalil
Department of Communication Engineering, Faculty of Electrical Engineering,
Universiti Teknologi Malaysia, 81310 Johor Bahru, Malaysia
e-mail: mdkamal@utm.my

N.A. Samsuri
e-mail: asmawati@utm.my

M.E. Jalil
e-mail: ezwanjalil@gmail.com

© Springer International Publishing Switzerland 2016
P.J. Soh et al. (eds.), *Theory and Applications of Applied Electromagnetics*,
Lecture Notes in Electrical Engineering 379, DOI 10.1007/978-3-319-30117-4_8

# 1   Introduction

Studies on wearable antennas have been rapidly increasing following the tremendous growth of body centric communication. Wearable antennas have been applied in medical, military, sports and in many other applications. Salonen et al. published the earliest study on wearable antenna in 1999 [1]. They proposed a dual-band planar inverted-F antenna using printed circuit board technology. Following that, studies on wearable antennas have been expanding including research on non-textile antennas as well as textile antennas. Some of the wearable non-textile antennas are designed in the form of printed circuit board [2], flexible material [3] and foam [4]. Textile antenna is literally the integration of antennas technology to the textile materials. The implementation of textile material as the substrate in wearable antennas provides conformity and flexibility on the human body. Textile antenna is appealing because it can be easily integrated into everyday's clothing [5]. Planar structure antenna is one of the most popular antennas to be used since it is suitable for many applications especially because of its small size. Therefore, this study will propose three types of planar textile antennas for body centric application. The proposed antennas will be designed, fabricated and tested to test their viability for body centric applications.

# 2   Textile Planar Antennas

## 2.1   Planar Straight Dipole

Initially, a planar straight dipole is designed due to its simplicity. Fleece is used as the substrate for the dipole. The fleece has permittivity $\varepsilon_r = 1.3$, tangent loss $\delta = 0.025$ and thickness $h = 1$ mm. Shieldit fabric is used for the conducting parts. To start with, the design parameters are numerically calculated based on established theoretical equations [6]. Subsequently, simulation is performed using CST Microwave Studio. The simulated dipole design is shown in Fig. 1a. The dipole's dimensions in mm are $L = 58$, $W = 14$, $L1 = 26.5$, $W1 = 6.5$ and $g = 1$ with overall size of $14 \times 58$ mm.

   Figure 1b shows the fabricated textile planar straight dipole. Pigtail SMA connector has been used in the prototype which is deemed suitable for wearable application compared to typical bottom-fed SMA connector. Figure 1c illustrates the comparison between measured and simulated return loss of the proposed dipole. From the graph, good agreement between measured and simulated $S_{11}$ can be observed. The measured return loss depth is $-14.7$ dB with bandwidth ranging from 2.24 to 2.62 GHz. The measured bandwidth of the textile planar straight dipole is 380 MHz (15.63 %). The measured and simulated radiation patterns of the textile planar dipole are presented in Fig. 2. From the radiation pattern plots,

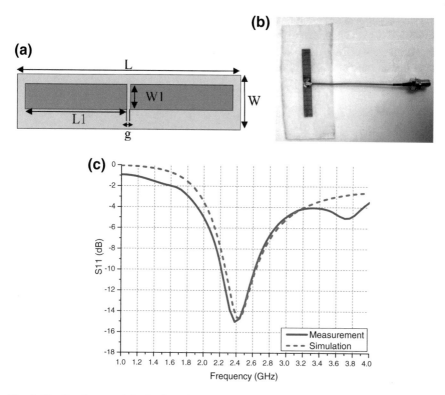

**Fig. 1** Textile planar straight dipole **a** simulated design, **b** fabricated prototype

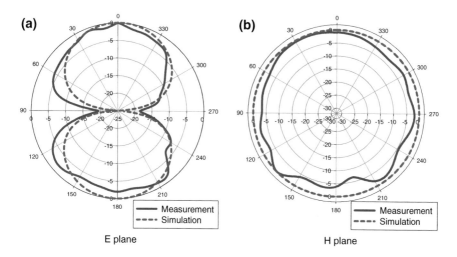

**Fig. 2** Radiation patterns of planar straight dipole at 2.45 GHz

measured patterns reasonably agree with the simulation giving the predicted eight shape for E plane and circle shape for H plane. The simulated gain is 1.77 dB, and measured gain is 2.93 dB. There are discrepancies between measured and simulated gain that is predicted due to dissimilarity between simulated and fabricated materials and structure.

## 2.2  Diamond Dipole

Next, diamond dipole is designed. Two diamond dipoles are designed to operate at 2.45 and 5.8 GHz, respectively. Diamond dipole is an inverted bow-tie dipole, and it offers broader bandwidth as opposed to conventional planar straight dipole. The parameters are initially calculated according to theoretical equations [7]. The dipole design layouts are as presented in Fig. 3a, b. The antennas are made of fleece and Shieldit fabric, similar as the previous design. The dimensions of the 2.45-GHz diamond dipole in mm are $L = 60$, $W = 45$, $L1 = 26.5$, $L2 = 18$, $W1 = 26.9$, $W2 = 3.4$ and $g = 1$. On the other hand, the 5.8 GHz has dimensions of $L = 35$, $W = 23$, $L1 = 11$, $L2 = 8$, $W1 = 14.4$, $W2 = 3.4$ and $g = 1$ in mm. Following the simulation process, the antennas are fabricated as shown in Fig. 3c, d. Pigtail SMA connector is used for both diamond dipoles similar as the previous design.

Figure 4a, b illustrates the comparison between measured and simulated result loss for both 2.45- and 5.8-GHz diamond dipoles, respectively. From the graphs, reasonable agreement between simulation and measurement is achieved for both dipoles. For the 2.45-GHz diamond dipole, the return loss depth is −33.64 and −15.39 dB for the 5.8-GHz dipole. The measured bandwidth for the 2.45-GHz dipole is 750 MHz ranging from 2.05 to 2.8 GHz. Meanwhile, the measured bandwidth for the 5.8-GHz dipole is 1.44 GHz ranging from 5.1 to 6.54 GHz. Slight discrepancies can be observed between measured and simulated return loss for both antennas. Compared to the simulated $S_{11}$, the measured results yield higher return loss depth. Such discrepancies are predicted due to dissimilarity between fabricated prototypes with simulated structures. To reduce the simulation time and size, the pigtail SMA connector is not included in the simulated structure. Discrete port has been used as the antenna excitation in the simulation. Figure 5 plots the measured and simulated radiation patterns for both 2.45- and 5.8-GHz textile diamond dipole.

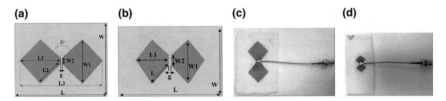

**Fig. 3** Textile diamond dipoles **a** 2.45-GHz simulated dipole design, **b** 5.8-GHz simulated dipole design, **c** 2.45-GHz fabricated antenna, **d** 5.8-GHz fabricated antenna

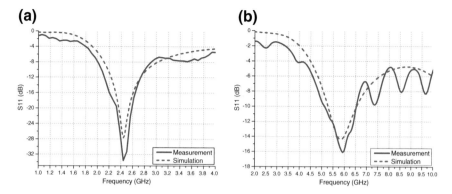

**Fig. 4** Measured and simulated $S_{11}$ of textile diamond dipoles **a** 2.45 GHz, **b** 5.8 GHz

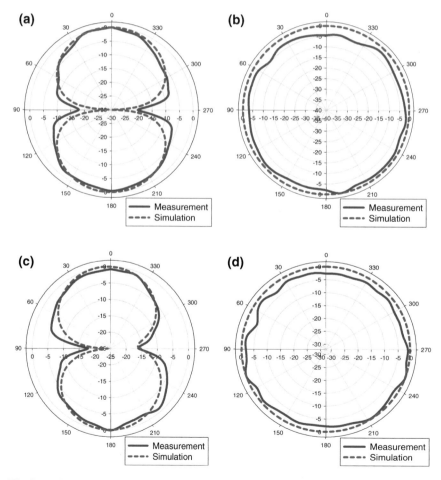

**Fig. 5** Radiation patterns of 2.45- and 5.8-GHz diamond dipoles **a** E plane—2.45 GHz, **b** H plane—2.45 GHz, **c** E plane—5.8 GHz, **d** H plane—5.8 GHz

From the plotted results, good agreement between simulation and measurements is achieved in E and H planes. The simulated and measured gains for 2.45-GHz diamond dipole are 2.04 and 3.09 dB, respectively. For the 5.8-GHz dipole, the simulated gain is 2.13 and 3.17 dB for the measured gain. Again, it is predicted that the discrepancies between measured and simulated gain are due to the dissimilarity between materials used in the simulation and measurement.

## 2.3  CPW Monopole

The final textile antenna that has been designed is CPW circular monopole, which is an ultra-wideband antenna. The antenna is a circular monopole that utilizes coplanar waveguide feeding technique. Circular monopole offers a very wide bandwidth. On the other hand, CPW feeding technique allows advantages such as high radiation efficiency, compact size and easy integration with system. Similar as the previous designs, the monopole is made of fleece and Shieldit fabrics. Initially, numerical calculation based on theoretical formulae is performed to estimate the antenna's parameters [8, 9]. Continuing from numerical calculation, simulation is conducted and the design layout of the monopole is shown in Fig. 6a. The dimensions of the CPW monopole in mm are $W = 50$, $L = 62$, $Wf = 4$, $g = 0.2$, $L1 = 15$, $W1 = 22.8$, $h = 0.025$, $R = 23$ and $s = 3$. Figure 6b shows the fabricated prototype of the textile CPW monopole. The measured and simulated return loss of the CPW monopole is presented in Fig. 6c. From the graph, it can be seen that there is slight discrepancy between simulation and measurement results. Nevertheless, similar trend is still observed between measured and simulated return loss. It is predicted that the discrepancy is resulted from dissimilarity between the simulated monopole and fabricated prototype. The measured bandwidth of the CPW mono-pole is 11.22 GHz. Figure 7 shows the radiation patterns of the fleece CPW monopole at 2.45 and 5.8 GHz. From the results, good agreement between simu-lated and measured patterns is observed at both 2.45 and 5.8 GHz. The measured gain at 2.45 GHz is 1.91 and 2.82 dB at 5.8 GHz. From the presented results, it can be seen that planar straight dipole acts as a narrowband antenna, diamond dipole as a wideband antenna and CPW monopole as an ultra-wideband antenna. Good agreement has been achieved between simulation and measurement. The proposed textile antennas can offer practicality and conformity for on-body communication.

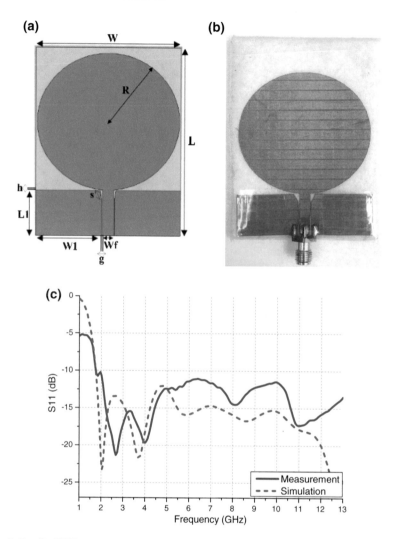

**Fig. 6** Textile CPW monopole **a** simulated design, **b** fabricated prototype, **c** $S_{11}$ results

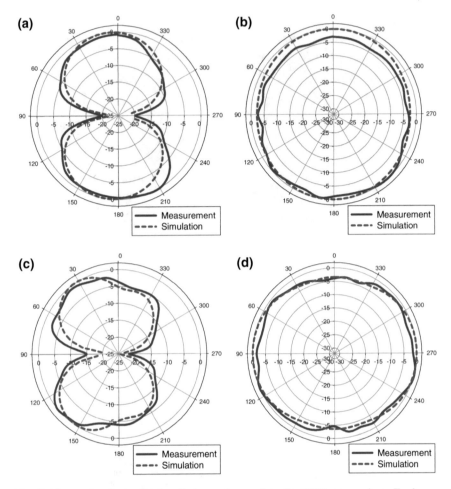

**Fig. 7** Measured and simulated radiation patterns of textile CPW monopole **a** E plane—2.45 GHz, **b** H plane—2.45 GHz, **c** E plane—5.8 GHz, **d** H plane—5.8 GHz

## 3 Wearable and Body Centric Measurements

### 3.1 Bending

Wearable antenna is prone to bending since it is exposed to body movement and posture. Bending measurement is necessary since it is challenging to retain textile antenna in its flat form. Polystyrene cylinder with permittivity of 1.06 is used as the bending gauge to demonstrate bending condition in human body. The polystyrene has diameter of 80 mm that mimics the size of human arm. Figure 8 shows bending measurement setup for planar straight dipole, diamond dipoles and CPW monopole. To fix the position of the antennas accordingly, thin sellotape is used.

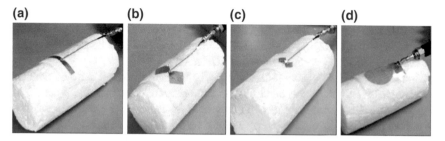

**Fig. 8** Planar textile antennas under bending measurement **a** planar straight dipole, **b** 2.45 GHz diamond dipole, **c** 5.8-GHz diamond dipole, **d** CPW monopole

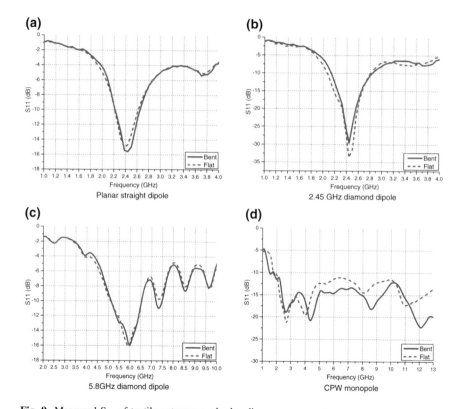

**Fig. 9** Measured $S_{11}$ of textile antennas under bending measurement

The measured return loss of bent versus flat textile antennas is shown in Fig. 9. From the results, good agreement between flat and bent curves is obtained for all antennas. Only small deviation can be observed between bent and flat $S_{11}$ results. For the case of planar straight dipole, the resonant frequency remains stable for the bending state with deeper return loss of −15.6 dB at 2.45 GHz. Slight increase in the impedance bandwidth is also observed. As for the 2.45-GHz diamond dipole, the return loss depth for the bent case is found to decrease to −29.6 from −33.6 dB for flat case. The measured bent of 5.8-GHz diamond dipole exhibits a slight shift of resonant frequency to 5.98 from 5.92 GHz.

For the case of CPW monopole, small deviation can be noticed between bent and flat antennas. However, the bend CPW monopole still preserves the wide 10 dB bandwidth, maintaining the wideband feature of the antenna. In general, return loss results remain stable for all antennas under bending test. Bending is found to exude minor effect to the impedance matching performance. The small size of the antennas under test in comparison with the size of the cylinder may be the factor to the negligible changes of the $S_{11}$. Figures 10, 11 and 12 illustrate the radiation pattern measurement of the bent textile planar straight dipole, 2.45- and 5.8-GHz diamond dipoles and CPW monopole, respectively. The polar plots compare the measured radiation patterns between bent and flat antennas. For all cases of antennas, good agreement can be seen between flat and bent results with minor deviation. However, from the results, it can be noticed that the beamwidths are slightly widened for all four antennas. Due to the bending curvature used in the bending experiment, the antennas are exposed to broader direction which leads the antennas to have slightly higher beamwidths.

As for the measured gain of the bent antennas, bent planar dipole exhibits gain of 2.7 dB compared to 2.93 dB gain of flat dipole. Small gain drop of 0.23 dB is observed for the bent case. Similar drop is also observed for the case of textile diamond dipoles. The measured gain of the bent 2.45-GHz diamond dipole is 2.63 dB, dropping from 3.09 dB for the flat case. The bent 5.8-GHz diamond dipole yields gain of 2.86 dB reducing from 3.17 dB gain of the flat antenna. As for the CPW monopole, the measured gain of the bent monopole is 1.83 dB dropping from 1.91 dB gain for the flat case. At 5.8 GHz, similar drop also occurs with measured

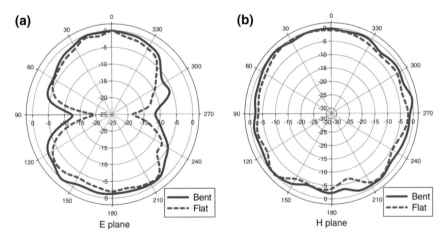

**Fig. 10** Measured radiation patterns of textile planar straight dipole under bending condition

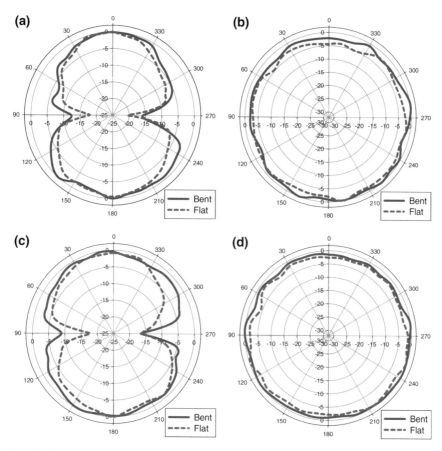

**Fig. 11** Measured radiation patterns of textile diamond dipoles under bending condition **a** E plane—2.45 GHz, **b** H plane—2.45 GHz, **c** E plane—5.8 GHz, **d** H plane—5.8 GHz

gain of bent monopole of 2.66 dB compared to 2.82 dB for the flat antenna. As mentioned earlier, bent antennas is exposed to wider direction leading them to become less directional, hence resulting to the slight gain drop. The wider beamwidths of bent antennas lead to the small gain reduction. In addition, the results show that the radiation pattern and gain at a higher frequency are slightly more sensitive to bending compared to lower frequency. Table 1 tabulates the comparison between all the proposed planar textile antennas in free space, in on-body environment and under bent condition. Results show that bending yields minor effect towards return loss, gain and radiation pattern performance. The resonant frequency and bandwidth are reasonably maintained even when the antennas are under bending exposure. Only small deviations in terms of slightly broader beamwidths are exhibited by the bent antennas. As a result, small drop of gain is observed for all the antennas under bending. Nonetheless, an acceptable agreement between flat and bend antennas has been obtained from the presented measured results.

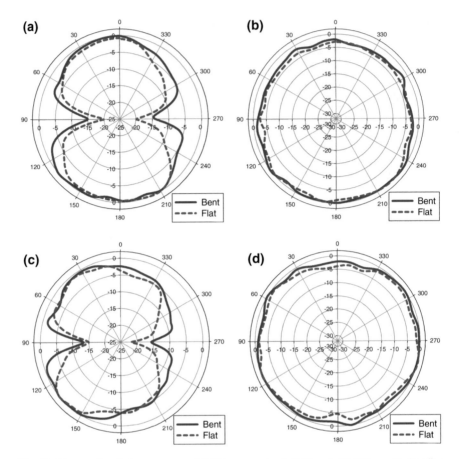

**Fig. 12** Measured radiation patterns of CPW monopole at 2.45 and 5.8 GHz under bending **a** E plane—2.45 GHz, **b** H plane—2.45 GHz, **c** E plane—5.8 GHz, **d** H plane—5.8 GHz

**Table 1** Comparison of planar textile antennas under body centric measurements at 2.45 GHz

| Antenna | $S_{11}$ | | | Gain (dB) | |
|---|---|---|---|---|---|
| | Free space | On body | Bent | Free space | Bent |
| Planar straight dipole | −14.7 | −13.6 | −15.6 | 2.93 | 2.7 |
| Diamond dipole | −33.64 | −13.44 | −29.6 | 3.09 | 2.63 |
| CPW monopole | −18.08 | −13.61 | −16.74 | 1.91 | 1.83 |

## 3.2 Wetness

Wearable antennas are exposed to wetness due to rain or sweat. Since the proposed planar textile antennas are made from textile materials, wetness test is necessary to be conducted. Wetness measurement has been taken for two antennas, i.e.

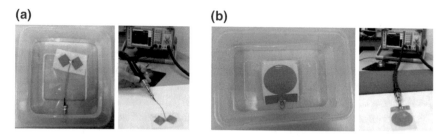

**Fig. 13** Wetness measurement for textile antennas **a** diamond dipole, **b** CPW monopole

2.45-GHz textile diamond dipole and CPW monopole. The wetness test involves observation at four different states of antenna, i.e. before washing, complete wet, damp and dry. Figure 13 shows the wetness measurement conducted for the two antennas. The antennas are soaked overnight for more than 12 h. The return loss of complete wet state antennas is measured immediately after being taken out from the water as shown in Fig. 13. Following that, the antennas are left to dry to measure the damp and completely dried antennas.

The measured return loss for both textile diamond dipole and CPW monopole under wetness test is shown in Fig. 14. From the results, it shows that the antennas suffer severe distortion under completely wet condition. Both of the antennas' performances have been influenced by the high permittivity of water. Under completely wet condition, the resonant frequencies are seen to be shifted to lower frequencies owing to the high permittivity of the wet antennas. The return loss curves of damp antennas are seen to closely return to the original curves. However, since there is still moisture left in the antennas, the return loss curves are not exactly similar with the original. On the other hand, the fully dried curves are observed to closely follow the original before washing curves. However, since the antennas are experiencing slight property changes as a result of shrinking, the exact original curve cannot be preserved.

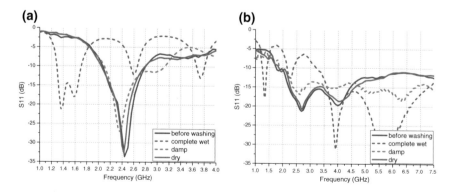

**Fig. 14** Measured $S_{11}$ of antennas under wet condition **a** diamond dipole, **b** CPW monopole

## 3.3   Specific Absorption Rate

Specific absorption rate (SAR) investigation is important for wearable antennas since they are placed very close to human body. SAR is a measure of radio frequency (RF) energy that is absorbed by a unit mass of body. The Federal Communications Commission (FCC) enforces SAR limit of 1.6 W/kg for a mass of 1 g, and International Commission on Non-Ionizing Radiation Protection (ICNIRP) imposes limit of 2 W/kg for 10 grams of tissues. SAR simulation is conducted using CST Microwave Studio that has inhomogeneous biological models called CST Voxel family. In the simulation, Gustav model, a 38-year-old male with 176 cm tall and 69 kg weight, was used. Gustav model has a complete organ model with conductivity, $\sigma$, and permittivity, $\varepsilon_r$, that comply with the standards tissue dielectric parameters recommended by FCC. In the SAR simulation, the antennas under test, i.e. 2.45-GHz diamond dipole and CPW monopole, are placed in the middle of torso part of Voxel biological body. Reference power of 1 W is excited, and the SAR value is investigated at 2.45 GHz.

Figure 15 illustrates 3D visualization of SAR values with information of maximum SAR position. From the SAR diagrams, it can be observed that high concentration of SAR values can be seen for both antennas when placed on the human body, respectively. From the diagram, the maximum 3D 10 g SAR values are 4.5 W/kg for textile diamond dipole and 2.37 W/kg for CPW monopole. On the other hand, the peak 1 g SAR for the diamond dipole is 9.29 W/kg, while CPW monopole has 4.32 W/kg peak 1 g SAR value. From the results, at 2.45 GHz, it can be observed that CPW monopole has lower peak SAR compared to diamond dipole. This is mainly due to the wideband nature of the CPW monopole antenna, and higher SAR value is expected to be at other frequency over the operating bandwidth.

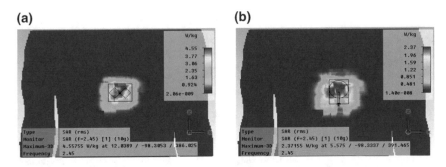

**Fig. 15** Simulated 3D maximum SAR visualization **a** diamond dipole, **b** CPW monopole

# 4 Conclusion

Three types of planar textile antennas, namely planar straight dipole, diamond dipole and CPW monopole, have been proposed in this study. All of the proposed antennas are designed, fabricated and measured. The antennas are made from entirely textile materials, with fleece and Shieldit fabrics as the substrate and conducting parts, respectively. The proposed antennas were tested under wearable and body centric measurements. Bending, wetness and SAR investigations were performed to test the antennas' performance for body communication realization. Bending was found not to yield any major performance deviation. On the other hand, since the antennas are made from non-waterproof textile, their performance is disrupted under wetness test. However, the original performance is retrieved once the antennas are dried out. SAR investigation shows slightly high values especially for the textile diamond dipole case. For future work, textile artificial magnetic conductor (AMC) ground plane is proposed to be included in order to reduce the SAR penetration into human body.

**Acknowledgements** The authors wish to thank Ministry of Higher Education (MOHE) and Universiti Teknologi Malaysia (UTM) for the Research Grants (Vote No: 02K02 and 4F277).

# References

1. Salonen P, Sydanheimo L, Keskilammi M, Kivikoski M (1999) A small planar inverted-F ant. for wearable app. In: The third international symposium on wearable computers, pp 95–100
2. Cibin C, Leuchtmann P, Gimersky M, Vahldieck R (2004) Modified E-shaped Pifa Ant. for wearable systems. In: URSI international symposium on electromagnetic theory, pp 873–875
3. Cibin C, Leuchtmann P, Gimersky M, Vahldieck R, Moscibroda S (2004) A flexible wearable antenna. In: IEEE international symposium on antennas and propagation, pp 3589–3592
4. Vallozzi L, Vandendriessche W, Rogier H, Hertleer C (2009) Design of a protective garment GPS ant. Microw Opt Technol Lett 51(6):1504–1508
5. Hall PS, Hao Y (2012) Antennas and propagation for body centric communications systems, 2nd Edn. Artech House, London, pp 63–64
6. Abu M (2012) Triple-band dipole ant. with AMC for radio frequency identification. Ph.D. dissertation. University of Teknologi Malaysia
7. Kaswiati WS, Suryana J (2012) Design and realization of planar bow-tie dipole array ant. with dual-polarization at 2.4 GHz freq. for Wi-Fi access point app. In: 7th international conference on telecommunication systems, services, and applications, pp 218–222
8. Balanis CA (2005) Antenna theory analysis and design. Wiley, New York
9. Wadell BC (1991) Transmission line design handbook. Artech House

# UHF-RFID Tag Antenna Using T-Matching and Double-Ended Rectangular Loop Techniques for Wristband Applications

**Tajchai Pumpoung, Pitchanun Wongsiritorn, Chuwong Phongcharoenpanich and Sompol Kosulvit**

**Abstract** This paper presents the UHF-RFID tag antenna applied on a human body (wrist). The proposed tag antenna was fabricated on a thin FR4 substrate ($\varepsilon_r = 4.3$) with the thickness of 0.25 mm. The T-matching and double-ended rectangular loop techniques are presented to improve an antenna impedance to achieve a matching condition with the NXP G2XL IC chip with the impedance of $21.29 - j191.7\ \Omega$, considering at the center frequency of the standard in Thailand with 922.5 MHz. The proposed tag antenna can apply for the standard of UHF-RFID in Thailand (920–925 MHz) and FCC (902–928 MHz). The radiation patterns of the tag antenna with and without wrist models are unidirectional and omnidirectional, respectively.

**Keywords** Tag antenna design · T-match · UHF-RFID tag antenna · Wrist

## 1 Introduction

Recently, the demand of the radio frequency identification (RFID) application is increased for using in many applications such as personnel access control, logistics, industrial process, product management, including identifying people, animals, and products [1]. The main application of ultrahigh-frequency (UHF) RFID is used for

T. Pumpoung (✉)
Faculty of Industrial Technology, Rambhai Barni Rajabhat University,
Chanthaburi 22000, Thailand
e-mail: t.pumpoung.1982@ieee.org

P. Wongsiritorn · C. Phongcharoenpanich · S. Kosulvit
Faculty of Engineering, King Mongkut's Institute of Technology Ladkrabang,
Bangkok 10520, Thailand
e-mail: kacher_911@hotmail.com

C. Phongcharoenpanich
e-mail: kpchuwon@kmitl.ac.th

S. Kosulvit
e-mail: kpsompol@kmitl.ac.th

© Springer International Publishing Switzerland 2016
P.J. Soh et al. (eds.), *Theory and Applications of Applied Electromagnetics*,
Lecture Notes in Electrical Engineering 379, DOI 10.1007/978-3-319-30117-4_9

product and commercial managements. However, the standard of the UHF-RFID in each country is different, for example the UHF-RFID standard of FCC-allocated frequency of the 902–928 MHz [2]. For Thailand, the allocated frequency of the UHF-RFID is 920–925 MHz [1]. One of the popular RFID applications is the human identification or body-worn application, in which the tag antenna is located on a human body. For wristband [3], the human skin has affected to the antenna performances due to high dielectric constant. The antenna performances such as gain, radiation pattern and impedance are changed.

In this paper, passive tag antennas for free space and a wristband application are designed, analyzed, and made up by using T-matching technique with double-ended rectangular loops. The proposed tag antennas are designed for NXP G2XL IC chip with the impedance of $21.29 - j191.7 \ \Omega$ at the frequency of 922.5 MHz [4]. An important requirement for the tag antenna design is that the tag antenna impedance must be conjugate-matched with the IC chip impedance to achieve a maximum power transfer from the tag antenna to the IC chip [5, 6].

Tag antenna design and analysis are described. The T-matching technique is one of the techniques, which can be used to improve the characteristic impedance of the antenna for matching with signal source (feed point). The research about the T-matching has been extensive in the literature [7–10].

## 2 The Tag Antenna Design

### 2.1 The Principle of the RFID System

The operating frequency bands of UHF-RFID systems can be divided into the frequency bands of 433/860/960 MHz [11] which depend on the standard of each country. The universal UHF-RFID frequency band is between 860 and 960 MHz. This frequency band covers the global UHF-RFID frequency band of the FCC standard with the frequency of 902–928 MHz and the Thai standard with the frequency of 920–925 MHz.

The effective isotropic radiated power ($EIRP_R$) is transmitted by the reader, and the sensibility of the tag transponder is $P_{IC}$. In Eq. (1), the read range of the tag assuming polarization is matched between the reader and the tag antennas is given in [6].

$$d_{max}(\theta, \phi) = \frac{c}{4\pi f} \sqrt{\frac{EIRP_R}{P_{IC}} \tau G_{tag}(\theta, \phi)}, \tag{1}$$

where $G_{tag}(\theta, \phi)$ is the tag gain.

$$\tau = \frac{4R_{IC}R_{Ant}}{|Z_{IC} + Z_{Ant}|^2} \leq 1 \tag{2}$$

$\tau$ is a power transmission coefficient which accounts for the impedance mismatch between the antenna ($Z_{Ant} = R_{Ant} + jZ_{Ant}$) and a IC chip ($Z_{IC} = R_{IC} + jZ_{IC}$) [8]. The tag antenna is designed to integrate with the IC chip with the impedance of $21.29 - j191.7 \ \Omega$, considering at 922.5 MHz. Therefore, the tag antenna impedance should be nearly $21.29 + j191.7 \ \Omega$ to meet the conjugate matching with the IC chip and to get a good efficiency.

The general matching technique to improve matching impedance of the tag antenna with the IC chip for UHF-RFID tag antenna is an inductively coupled loop and a T-matching as presented in the recent open scientific literature [6]. To analyze the characteristic of the antenna, the tag antenna and equivalent circuit are presented in Fig. 1. Each part of T-matching relates to a capacitance and an inductance that exactly affect the antenna reactance. A folding dipole is related to a capacitance, an inductance, and a resistance that have the effect on both of the antenna resistance and of the reactance.

To simplify the antenna parameters, the size of the initial antenna is defined as follows: The width of the tag antenna ($w_1$) is $b + 2g$ when $b$ is 8 mm. The total antenna length ($l_1$) is around $\lambda + 2g$ or 161 mm. The gap ($g$) between the substrate edge and the antenna radiation part is 1 mm. The $w$ and $w'$ are fixed at 1 mm which is the width of the folding dipole and the T-matching line. The $C_i, C_o$ and $C_w$ are the parameters of the IC chip platform with the size of 3, 7, and 4 mm, respectively. The parameter $a$ is the width of T-matching which is fixed at 30 mm, and the folded dipole the length ($l$) is around $\lambda$ or 159 mm.

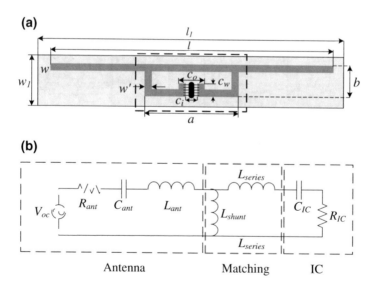

**Fig. 1** The structure (**a**) and equivalent circuit (**b**) of antenna with T-matching [11]

## 2.2 Parametric Studies and Improvement of the Antenna Impedance on Free Space

The first parameter that is the length $a$ is varied. Figure 2 are the tag impedance for various parameter $a$. It can be seen that the length of $a$ has direct variation with the reactance until $a = 39$ mm. After that, the tendency of the reactance will be reduced, considering at the frequency 922.5 MHz (center frequency of Thailand standard). The initial antenna at 922.5 MHz has the impedance close to the IC chip when $a = 39$ mm with $133.46 + j187.53$ $\Omega$. Obviously, by varying the parameter a, the reactant of the tag antenna is close to the desired reactance.

The initial tag antenna resistance is reduced in this second step, by assuming the length folding dipole is related to the resistance; therefore, the folding dipole is examined to search for the appropriate tag antenna resistance.

From the previous step, the antenna reactance is reached to the proposed reactance with $133.46$ $\Omega$; however, the resistance is $133.46$ $\Omega$, which is fairly high. Then, the rectangular loop technique is presented to decrease the initial tag antenna resistance. The width of the ended rectangular loop ($b$) is 8 mm. The initial ended rectangular loop length ($s_l$) is equal to the total folding dipole length ($l$) of $\lambda$ or 159 mm. The UHF-RFID tag antenna with T-matching and double-ended rectangular loops is shown in Fig. 3.

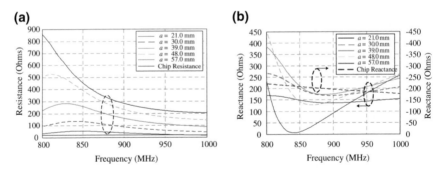

**Fig. 2** The resistance (**a**) and the reactance for various parameters $a$ (the initial tag antenna)

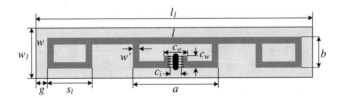

**Fig. 3** The UHF-RFID tag antenna with T-matching and double-ended rectangular loops

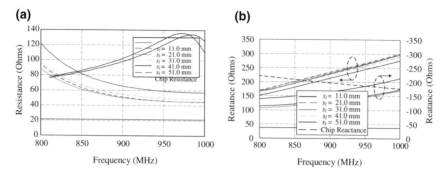

**Fig. 4** The resistance (**a**) and the reactance (**b**) for various parameters $S_l$ for improving the impedance of the UHF-RFID tag antenna with T-matching and double-ended rectangular loops

From Fig. 4, it is shown that the resistance of the square end loop tag antenna is decreased when the parameter $s_l$ is increased. The antenna resistance can be decreased to the lowest point when $s_l = 31$ mm with the impedance of $47.67 + j243.04\ \Omega$.

Although the tag antenna resistance is decreased from the previous step, the resistance and reactance of the designed tag antenna need to be improved to earn the proper impedance matching with the desired IC chip because in this step, the resistance and reactance are higher than those of the IC chip.

In the third step, parameter $a$ is varied again to improve the matching condition and to reduce the designed tag antenna's resistance and reactance.

Figure 5 shows the impedance and the $|S_{11}|$ (dB) of the antenna by varying parameter $a$ again. It can be seen that when the parameter $a = 28$ mm, the desired matching condition is obtained with the impedance and $|S_{11}|$ (dB) of the tag antenna designed with $20.75 + j188.72\ \Omega$ and $-22.92$ dB, respectively, considering at the frequency of 922.5 MHz.

In the final step, parameter $s_l$ is varied again to achieve the best matching impedance. At $s_l = 17$, the proposed tag antenna has the appropriate matching condition with the desired IC chip with $21.69 + j189.16\ \Omega$ with the $|S_{11}|$ (dB) of $-24.54$ dB as shown in Fig. 6. It is obvious that the matching condition is improved in this step.

## 2.3 Parametric Study and Improvement of the Antenna Impedance Applied for Wrist

In this section, the antenna applied on a wrist is designed and studied based on the antenna structure in Sect. 2.2. Although the antenna in Sect. 2.2 can operate properly at the frequency band of UHF-RFID in Thailand according to the FCC standard, the high dielectric constant of the human skin ($\varepsilon_r = 41.29$ and loss tangent $= 0.874$ [3]) can affect the tag antenna characteristics such as shifting the

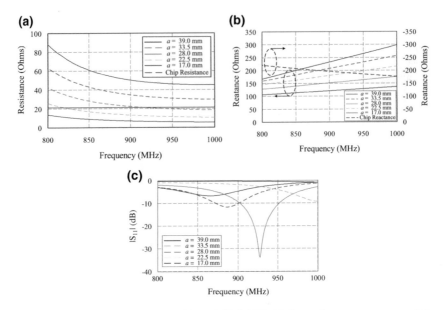

**Fig. 5** The resistance (**a**), reactance (**b**), and $|S_{11}|$ (dB) (**c**) for various parameters $a$ of the UHF-RFID tag antenna with T-matching and double-ended rectangular loops

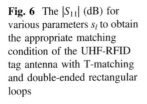

**Fig. 6** The $|S_{11}|$ (dB) for various parameters $s_l$ to obtain the appropriate matching condition of the UHF-RFID tag antenna with T-matching and double-ended rectangular loops

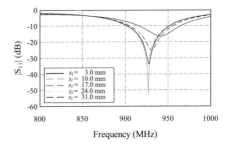

antenna impedance, the resonance, the gain, and the radiation pattern. Therefore, in this section, the antenna is redesigned for applying on wrist. Besides, for the proposed application (wristband), the parametric study and the effect of the curved tag are presented and conducted in this section. The initial tag structure is the final tag structure in Sect. 2.2, which is designed using wrist model as shown in Fig. 7a. The human wrist model has the diameter of 70 mm.

The curvature of the tag antenna in Sect. 2.2 ($l = 159$ mm) is applied on the wrist because it has affected to the resonant frequency. The resonant frequency is shifted to the lower frequency as shown in Fig. 7b. This impact is caused by the high dielectric constant of the human skin; therefore, when the tag is located on the high dielectric object, the electrical length of the antenna is increased. To improve the impedance matching, the antenna length needs to be decreased. As shown in

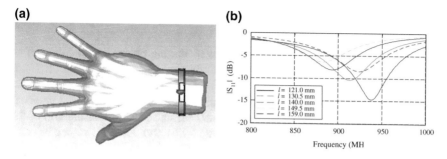

**Fig. 7** The wrist model [12] with the proposed tag antenna (**a**) and the $|S_{11}|$ (dB) for various parameters $l$ (**b**)

Fig. 7b, at $l = 121.0$ mm, the antenna operating frequency mostly covers the frequency band of FCC standard with the $|S_{11}|$ (dB) at the frequency band of 922.5 MHz of $-11.16$ dB.

Nevertheless, the antenna parameter is varied to obtain the appropriate matching condition when the tag is attached to the wrist. Then, the parameter $a$ is varied to improve matching condition as explained in Fig. 8a.

When $a = 30$ mm, the $|S_{11}|$ (dB) covers the operating frequency of FCC and Thailand UHF-RFID standard. In Fig. 8b, parameter $s_l$, is varied, and at $s_l = 21$ mm, the $|S_{11}|$ (dB) with $-35.4$ dB at 913 MHz is obtained.

The final step is to improve matching condition of the antenna designed for wrist model. It is found that the best parameter can be achieved when $l = 122$ mm as shown in Fig. 9. The $|S_{11}|$ (dB) covers the frequency band of 895–941 MHz that in turn covers the desired operating frequency of FCC and the standard in Thailand.

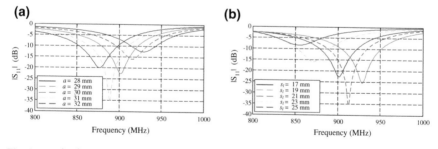

**Fig. 8** The $|S_{11}|$ (dB) for various parameters $a$ (**a**) and $S_l$ (**b**) the antenna designed for wrist

**Fig. 9** The $|S_{11}|$ (dB) for various parameters $l$ of the antenna designed for wrist in final step

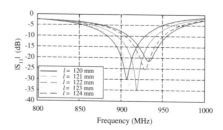

## 3   Antenna Characteristics

Figure 10a shows the simulated impedance of the tag antenna with and without wrist models. The impedance, considering at the frequency band of 922.5 MHz, is $21.69 + j189.16\ \Omega$ (without wrist model) and $21.70 + j193.84\ \Omega$ (with wrist model), respectively. As Fig. 10b shows, the $|S_{11}|$ (dB) of the proposed antennas with and without wrist are $-24.54$ and $-25.97$ dB and percent transmission of the proposed antennas with and without wrist are 99.64 and 99.74 %, respectively.

### 3.1   The Impedance Characteristics

See Fig. 10.

### 3.2   The Radiation Pattern and Gain of the Antenna

The patterns of the tag antenna are the omnidirectional beam (without wrist model) and unidirectional beam (with wrist model) as shown in Fig. 11. Figure 12 shows the gain of the proposed tag antenna with 2.42 dBi (without wrist model) and $-11.21$ dBi (with wrist model), considering at the frequency band of 922.5 MHz (center frequency of Thailand).

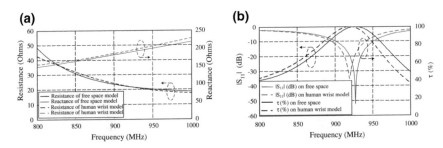

**Fig. 10 a** The impedances and **b** the $|S_{11}|$ (dB) and the percent power transmissions of the UHF-RFID tag antenna with T-matching and double-ended rectangular loops (with and without wrist model)

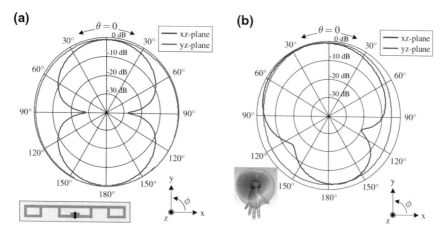

**Fig. 11** The radiation patterns, **a** without wrist model and **b** with wrist model

**Fig. 12** The gain of the RFID tag antenna with T-matching and double-ended rectangular loops on free space and human wrist model

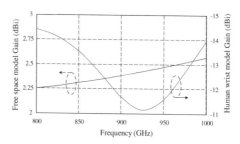

## 4 Fabrications and Measurement

The prototype antennas were fabricated by using the parameters in the final step of the tag antenna design in the Sects. 2.2 and 2.3. The proposed tag antennas on free space and wrist model have the size of $161 \times 10$ mm$^2$ and $220 \times 10$ mm$^2$, respectively. The photograph of the proposed antenna is depicted in Fig. 13. The maximum read range was measured by using Symbol Reader model XR480 operating at the frequency band of FCC (902–928 MHz) (4 W EIRP) with a linearly polarized antenna. The obtained maximum read ranges of the proposed tags without and with wrist are 8.30 and 2.50 m, respectively.

**(a)**

**(b)**

**(c)**

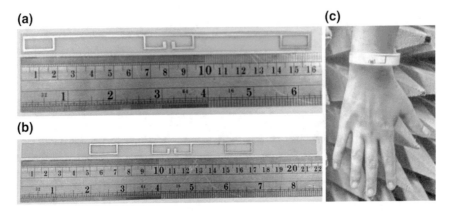

**Fig. 13** The photograph of the prototyped tag antenna, **a** free space model, **b** human wrist model, and **c** wristband from human wrist model

## 5   Conclusions

The tag antenna for the UHF-RFID system in Thailand is presented in this paper. The proposed tag antenna can be used properly for the UHF-RFID of Thailand standard and the FCC standard. The antennas, which are designed in free space and on wrist model, have the dimensions $161 \times 10$ mm$^2$ and $220 \times 10$ mm$^2$, respectively. The proposed antennas can be achieved by using the T-matching and the double-ended rectangular loops to improve the impedance matching. The T-matching technique is added to improve the impedance matching of the tag antenna. The appropriate input impedance of the tag antenna for the conjugate matching with the desired IC chip impedance is found to be $21.29 + j191.7$ $\Omega$. The impedance of the proposed tag antenna without wrist is $21.69 + j189.16$ $\Omega$ considering at 922.5 MHz with the power transmission of 99.64 %. The impedance of the proposed tag antenna which is applied on wrist is $21.70 + j193.84$ $\Omega$ with the power transmission of 99.74 %. The radiation patterns are the omnidirectional beam without wrist model and unidirectional with wrist model. The maximum read ranges of the proposed antennas are 8.30 m (without wrist model) and 2.5 m (with wrist model). This tag antenna is appropriated for the UHF-RFID system in Thailand.

## References

1. Pumpoung T, Phongcharoenpanich C (2015) Design of wideband tag antenna for ultra-high-frequency radio frequency identification system using modified t-match and meander-line techniques. Electromagnetics 35(5):340–354
2. FCC Part 15.247 Rules Systems Using Digital Modulation. http://www.semtech.com/images/datasheet/fcc_digital_modlation_systems_semtech.pdf

3. FCC: Body Tissue Dielectric Parameters. https://www.fcc.gov/encyclopedia/body-tissue-dielectric-parameters
4. NXP Semiconductors. http://www.nxp.com
5. Pozar DM (2008) Microwave engineering, 3nd edn. USA
6. Marrocco G (2008) The art of UHF RFID antenna design: impedance matching and size-reduction techniques. IEEE Antennas Propag Mag 50(1):66–79
7. Balanis CA (2005) Antenna theory. Analysis and design, 3nd edn. Hoboken
8. Mohhammed NA, Demarest KR, Deavours DD (2010) Analysis and synthesis of UHF RFID antennas using the embedded T-match. In: Proceedings of the IEEE international conference on RFID 2010. IEEE Press, Orlando, FL, pp 230–236
9. Nohuchi K, Mizusawa T, Betsudan SI, Makino S, Sasaki T (2009) Broadband matching of a RFID antenna by using a T-type circuit. In: Proceedings of the IEEE international workshop on antenna technology 2009. Santa Monica, CA, pp 1–4
10. Choo J, Ryoo J, Hong J, Jeon H, Choi C, Tentzeris MM (2009) T-matching networks for the efficient matching of practical RFID tags. In: Proceedings of the European microwave conference, IEEE Press, Rome, pp 5–8
11. Dobkin DM (2008) The RF in RFID passive UHF in practice. Burlington
12. Realistic Hand (2015) http://tf3dm.com/3d-model/freerealsichand-85561.html

# Low-Profile Metallic Tag Antenna for UHF-RFID System

Kittima Lertsakwimarn and Chuwong Phongcharoenpanich

**Abstract** A low-profile ultrahigh frequency (UHF) passive radio frequency identification (RFID) tag for the applications to distinguish metallic from non-metallic is presented. The proposed antenna structure consists of T-shaped structure and parasitic parallel longitudinal strip lines which are placed symmetrically. There is an air gap between the head parts of T-shaped structure for installing IC chip. The head part has shorting pin that connects the ground plane. The parametric studies are described. From the results, the tag antenna with defected ground plane has larger difference of maximum read range compared between the tag on metallic plate and the tag in free space. The proposed UHF-RFID tag antenna can adjust the impedance characteristic by changing the length of the middle longitudinal strip. The tag antenna prototype is fabricated with an area of $90 \times 20 \times 3$ mm$^3$ ($0.28\lambda_0 \times 0.06\lambda_0 \times 0.009\lambda_0$). The measured results of maximum read range on the metallic plate and in free space are 83 and 24 mm, respectively. The simulated and measured results of the proposed RFID tag antenna are discussed.

**Keywords** Low-profile tag antenna · UHF-RFID tag antenna · Metallic tag antenna

## 1 Introduction

The current radio frequency identification (RFID) is a technology used for identifying objects. There are many applications of RFID technology, such as tracking, warehouse, retail, manufacturing, and supply chain automation. The application

K. Lertsakwimarn (✉)
Faculty of Industrial Technology, Rambhai Barni Rajabhat University,
Chanthaburi 22000, Thailand
e-mail: wimand.mai@gmail.com

C. Phongcharoenpanich
Faculty of Engineering, King Mongkut's Institute of Technology
Ladkrabang, Bangkok 10520, Thailand
e-mail: kpchuwon@kmitl.ac.th

© Springer International Publishing Switzerland 2016
P.J. Soh et al. (eds.), *Theory and Applications of Applied Electromagnetics*,
Lecture Notes in Electrical Engineering 379, DOI 10.1007/978-3-319-30117-4_10

109

continues with the announcement in a very wide area [1] because it has better ability than barcodes and infrared technology. Various bands in RFID systems, such as low frequency (LF), high frequency (HF), ultrahigh frequency (UHF), and microwave frequency, are very popular because the operating range is greater than the LF- and HF-RFID systems. The LF- and HF-RFID systems use the near-field coupling techniques, whereas the UHF-RFID utilizes the far-field radiation technique. The RFID system operation consists of tag, reader, and information management system. Antenna reader transmits radio energy to investigate the tag antenna attached to the product to be identified.

Generally, the types of tag antenna can be classified into three different categories: active, semi-active, and passive. The active and semi-active tags require a battery, and it will limit the time of life. Passive tags have no problems with energy restrictions. Therefore, passive RFID system is used in volume applications, such as warehouses and retail stores [2–4]. In supply chain system, there are several products made from different materials such as wood, plastic, and metal. At the same time, the tag is sensitive to the performance when attached to a different material, especially in metallic material. There is a limitation in the usage of tag antenna with metallic material. The metallic objects make strong reflection of the electromagnetic waves in RFID communication. The performance of most RFID tags will be degraded when it is mounted with metallic object. Therefore, in the design of tag antenna that works well with the metallic material, it is an interesting topic. One of the most challenges in the design of RFID tag is to reduce the effects of interference from the metallic surface, which is achieved, for instance, by inserting the tag in a high-permittivity substrate or embedding it with ground plane [2].

In recent years, there are many researches that paid attention to a design of metal RFID tag antenna. Many antenna types have been proposed such as the invert-F antenna (IFA) [5], planar invert-F antenna (PIFA) [6], patch-type antenna [7], loop antenna [8], low-profile RFID tag antenna using compact AMC substrate [9], and antenna with HIS ground plane [10].

This paper proposes the low-profile RFID tag antenna for the applications to distinguish metallic from non-metallic objects. The allocated UHF band in Thailand is from 920 to 925 MHz. In Sect. 2, the tag antenna configuration is shown. The antenna evolution and simulated results of the proposed antennas are given in Sect. 3. Section 4 presents our measured results. Finally, conclusion is discussed in Sect. 5.

## 2 Antenna Configuration

The design antenna consists of T-shaped structure and parasitic parallel longitudinal strip lines on the one side of a FR-4 substrate ($\varepsilon_r = 4.3$) and defected ground plane on the other side as shown in Fig. 1. The antenna is designed with the thickness $h$ of 3 mm. The advantages of the designed structure include low profile, lightweight, ease of fabrication, and low fabrication costs.

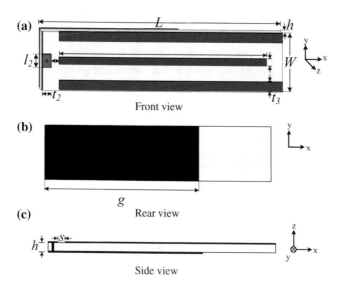

**Fig. 1** The proposed antenna configuration

**Table 1** Physical dimension of the antenna parameters

| Parameters | $L$ | $W$ | $l_1$ | $l_2$ | $t_1$ | $t_2$ | $t_3$ | $s$ | $g$ |
|---|---|---|---|---|---|---|---|---|---|
| Size (mm) | 90 | 20 | 75.4 | 4 | 2 | 3.14 | 5.5 | 2 | 50 |

As shown in Fig. 1a, the parasitic parallel longitudinal strip lines have symmetry in y-axis of the tag antenna. An air gap between the head and footer parts of T-shaped structure ($S$) is 2 mm. A shorting pin is connected to the head part of T-shaped structure and the ground plane as shown in Fig. 1c. The length of the ground plane ($g$) is defected by choosing $g = 50$ mm to provide maximum performance when used with metallic materials. The overall size of the tag antenna ($L \times W \times h$) is $90 \times 20 \times 3$ mm$^3$. The width ($W$) and length ($L$) of the tag antenna are suitable for implementation, which is convenient to mount and remove tag antenna with mountable objects. The optimal parameters are shown in Table 1.

# 3   Antenna Design and Simulation

In this section, the antenna design and the simulated results will be addressed. The antenna was designed using the CST MW Studio [11]. The aim of the proposed tag antenna is designed to work well with metallic objects. In other words, this tag antenna does not work when it is attached to other material such as air, plastic and etc. Therefore, the metallic plate will be installed behind the tag antenna. The metallic plate size is $1.5\lambda \times 1.5\lambda$ mm$^2$. The antenna evolution is initialized with

T-shaped to study the impedance characteristic variations. Next, the second will be added to parasitic parallel lines into above and below the T-shaped structure along x-axis. The length of the T-shaped structure and width of the parasitic parallel lines can be used for adjusting the resistance and the reactance of tag antenna to meet the required specifications. Finally, the effect of the ground plane size will be studied by reducing the ground plane size.

### 3.1  T-Shaped Structure

Most of the tag antennas are designed based on conjugate matching with the chip impedance to obtain maximum power transfer. It will enhance performance, gain, radiation efficiency, and polarization of tag antenna. In this paper, metal tag antenna is designed for UCODE G2XL chip [12], whose input impedance characteristic is about $21.65 - j191.70\ \Omega$ at center frequency (922.5 MHz). The evolution of tag antenna will start with T-shaped structure on the one side of the dielectric substrate. There is an air gap ($S$) between the head and footer of T-shaped structure for installing the IC chip as shown in Fig. 2a. At the center of head of T-shaped structure is installed shorting pin to connect the ground plane. The length of the footer of T-shaped structure ($l_1$) is initialized with $\lambda/2$. Variation in the length of the T-shaped structure can be used for adjusting the resistance and the reactance of tag antenna to meet the required specifications.

The increased parameter $l_1$ (length of the longitudinal strip) is also studied. The impedance of these parameter variations at the operating frequency of 922.5 MHz is shown in Fig. 3. An increase in parameter $l_1$ from 30 to 80 mm leads to a gradual increase in the reactance and the resistance. Therefore, the parameter $l_1$ can control the impedance of the RFID tag antenna. However, the adjusted length of the longitudinal strip of T-shaped structure is not appropriate to adjust the RFID tag antenna to match the impedance well. Next, parasitic parallel lines are installed above and below the T-shaped structure.

**Fig. 2** T-shaped structure antenna configuration

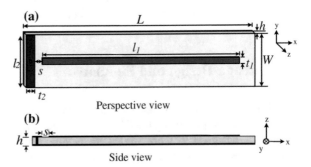

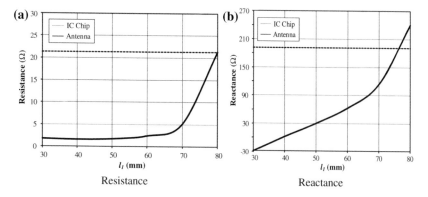

**Fig. 3** Simulated impedance versus $l_1$ at the operating frequency on metallic plate

## 3.2 T-Shaped Structure and Parasitic Parallel Line Antenna

The parasitic parallel lines are placed symmetrically parallel to the longitudinal strip of the T-shaped structure. The length of the parasitic parallel lines starts at the same location as the length of the longitudinal strip of the T-shaped structure until the end edge of the structure as shown in Fig. 1a. The impedance characteristic of the RFID tag antenna will be studied by increasing parameter $t_3$ (width of the parasitic parallel lines). Increasing the width will increase the width of the parasitic parallel lines from the edge into inside of the tag antenna.

Figure 4 shows the effect of parasitic parallel lines on the impedance characteristic when $l_1 = 75$ mm and $l_2 = 4$ mm. As a result, when $t_3$ is increased from 0 mm to 4 mm at the operating frequency of 922.5 MHz, the impedance is gradually increased to match the impedance well. Therefore, the impedance of the RFID tag antenna can conjugate-match by adjusting the longitudinal strip length ($l_1$) and width of the line parasitic parallel ($t_3$).

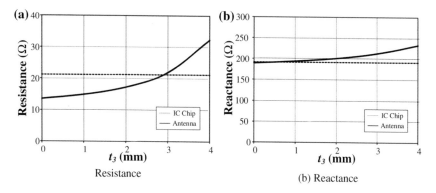

**Fig. 4** Simulated impedance versus $t_3$ at the operating frequency on metallic plate

### 3.3 T-Shaped Structure and Parasitic Parallel Antenna with Defected Ground Plane

In Sect. 3.2, we have designed the T-shaped structure and parasitic parallel antenna. The tag antenna shows good conjugate matching with IC chip when it is attached to the metallic plate. When calculating the percentage of transmission coefficient, it is changed from 96.8 to 42.2 % when it is mounted on a metallic plate and in a free space, respectively. The impact of the defected ground plane size on the maximum read range will be discussed in this section. The maximum read range can be calculated by the Friis free space transmission formula as

$$r_{max} = \frac{\lambda}{4\pi} \sqrt{\frac{P_t G_t G_r (1 - \Gamma^2)}{P_{th}}} \tag{1}$$

where $P_t$ is the power transmitted by the RFID reader, $G_t$ is the gain of the transmitting antenna ($P_t G_t$ is EIRP (equivalent isotropic radiated power)), $G_r$ is the gain of the receiving tag antenna, and $P_{th}$ is the minimum threshold power necessary to power up the IC chip. The percentage difference of the maximum read range of the tag antenna with $g = 90$ mm was mounted on a metallic plate and placed in free space. As a result, it is equal to 50 %, which is less than desirable.

Figure 5 shows the maximum read range of the tag antenna when attached in free space and on metallic plate versus ground plane. The maximum read range of the antenna when attached on metallic plate is relatively stable and it has changed when placed in a free space. Therefore, deflecting the ground plane will affect the reading range of the tag antenna when placed in free space. The percentage difference of maximum read range between mounted on metallic plate and in free space is shown in Fig. 5. It is found that defected ground plane only of 50 mm has a maximum percentage value. The maximum percentage difference of maximum read range is equal to 95 %.

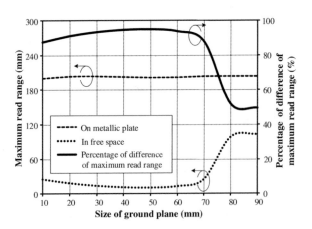

**Fig. 5** Calculated result of read range antenna when in free space and on metallic plate versus ground plane

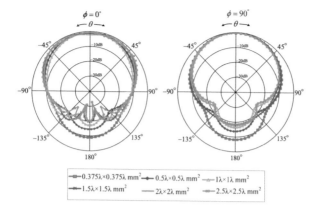

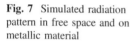

**Fig. 6** Simulated radiation patterns with different sizes of metallic plate

**Fig. 7** Simulated radiation
pattern in free space and on
metallic material

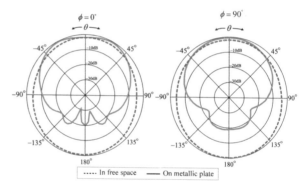

The simulated result of radiation pattern in XZ-plane and YZ-plane is shown in Figs. 6 and 7. In Fig. 6, it is shown that the direction of the main beam of the antenna does not change according to the size of the metallic plate. The simulated result of radiation pattern of tag antenna on $1.5\lambda \times 1.5\lambda$ metallic plate and in free space at the operating frequency is shown in Fig. 7. The half-power beam width in XZ-plane and YZ-plane is about $63.8°$ and $70.4°$, respectively. The front-to-back ratio is 12.7 dB.

As shown in Table 1, these parameters can achieve the impedance of $22.46 + j197.14 \ \Omega$ and $3.6 + j93.54 \ \Omega$, respectively considering at the operating frequency (922.5 MHz). The $|S_{11}|$ of the proposed antenna on metallic plate is less than $-3$ dB that covers the operating frequency of 914–926 MHz. It is noted that this frequency range can cover the operating frequency of UHF-RFID in Thailand (920–925 MHz) (Fig. 8).

**Fig. 8** Simulated $|S_{11}|$ of tag antenna on metallic plate

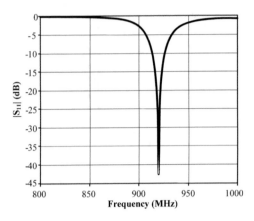

## 4 Measured Result

The antenna is measured in Anechoic chambers which is all reflected waves from nearby objects and the ground are suppressed. RFID Reader model is chooses in Motorola XR450 [13] and the standard dipole antenna of reader model is MP651B. The linearly polarized reader antenna with the gain of 2.15 dBi [14] was used in the measurements. The photograph of the prototype of the proposed tag antenna is depicted in Fig. 9. The tag antenna was measured in free space and attached to a 1.5λ × 1.5λ aluminum metallic plate, respectively. The maximum read range of tag antenna in free space and on metallic plate are 24 and 83 mm, respectively.

**Fig. 9** Photograph of the prototype antenna

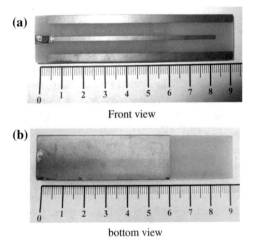

# 5  Conclusions

This paper has presented a low-profile UHF-RFID tag antenna which can distinguish metallic from non-metallic. The structure consists of T-shaped structure and parasitic parallel longitudinal strip lines. The impedance characteristic of the proposed antenna can be controlled by adjusting the length of the T-shaped structure and width of the parasitic parallel lines. The effect of the defected ground plane has been discussed. As a result, the defected ground plane $g$ of 50 mm can achieve the largest difference of maximum read range compared between the tag on metallic plate and the tag in free space. The simulated results illustrated the performance of the proposed antenna in terms of $|S_{11}|$, and the advantage is that the proposed antenna is suitable for fabrication. The obtained result shows that the maximum read range of tag antenna in free space and on metallic object when attached to a $1.5\lambda \times 1.5\lambda$ aluminum metallic plate is 24 and 83 mm, respectively. The proposed antenna can work well with metallic products; therefore, it can reduce the impact of metals on the efficiency of the RFID tag antenna.

# References

1. Chen SL, Kuo SK, Lin CT (2009) A metallic RFID tag design steel bar and wire-rod management application in the steel in-dustry. Prog Electromagn Res PIER 91:195–212
2. Sydanheimon L, Ukkonen L, Kivikoski M (2006) Effects of size and shape of metallic objects on performance of passive radio frequency identification. Int J Adv Manuf Technol 30:897–905
3. Finkenzeller K (2003) RFID Handbook, 2nd ed. Wiley, New York
4. Raza N, Bradshaw V, Hague M (1999) Application of RFID technology. IEE Colloquium RFID Technol 123:1/1-1/5
5. Kim KH, Song JG, Kim DH, Hu HS, Park JH (2007) Fork-shaped RFID tag antenna mountable on metallic surfaces. Electron Lett 43(25):1400–1402
6. Kwon H, Lee B (2005) Compact slotted planar inverted-F RFID tag mountable on metallic objects. Electron Lett 41(24):1308–1310
7. Yu B, Kim SJ, Jung B, Harackiewicz FJ, Lee B (2007) RFID tag antenna using two-shorted microstrip patches mountable on metallic objects. Microw Optical Technol Lett 49(2):414–416
8. Lin D-B, Wang C-C, Chou J-H, Tang I-T (2012) Novel UHF RFID loop antenna with interdigital coupled section on metallic objects. J Electromagn Waves Appl 26(2–3):366–378
9. Kim D, Yeo J (2008) Low-profile RFID tag antenna using compact AMC substrate for metallic objects. IEEE Antennas Wirel Propag Lett 7:718–720
10. Liu Y, Luk K-M, Yin H-C (2010) A RFID tag metal antenna on a compact HIS substrate. Prog Electromagn Res Lett 18:51–59
11. CST-Microwave Studio (2006) User's Manual
12. NXP, Ultrahigh frequency smart label ICs [online] http://www.nxp.com/acrobat_download/literature/9397/75016225.pdf
13. http://www.motorola.com/Business/US-EN/Business+Product+and+Services/RFID/RFID+Readers/XR450+Fixed+RFID+Reader_US-En
14. http://www.testmart.com/sp.cfm/ANT/ANRI/MP651B.html

# Multiband Antenna for Robotic Swarming with Application of Dynamic Boundary Tracking

Malay Ranjan Tripathy, Priya Ranjan and Arun Kumar Singh

**Abstract** Dynamic swarming of robots requires equally aggressive communication capability even in the event of dynamic network formations. "On-the-fly" communication can pull off real surprises in a matter of microseconds. This motivates a new class of multiband antenna design to accommodate various classes of robotic platforms and to enhance their communication capability. To address this need, we propose an antenna which is designed on FR4 substrate with a thickness of 1.6 mm, $\varepsilon_r = 4.4$, and loss tangent 0.02. The dimension of the antenna is $35 \times 30 \times 1.6$ mm$^3$. Microstrip-line feeding is used in this antenna. It has a structure which is a combination of split ring resonator and meander line resonator. This design is simple, compact, and miniaturized. It seems promising to integrate with the rest of the dynamic boundary tracking robotic circuit.

**Keywords** Robotics · Multiband antenna · Swarming communication · IoT · White space · ISM

## 1 Introduction

Ours is the era of distributed robotics. Dynamic boundary tracking has been a hot area of research activity due to affordable communication, sensing, and mobile platform/robotic technologies. While large amount of attention has been paid toward swarming control scheme design, sensors, and communication paradigm at transport and network layers, there are issues to be dealt with at physical layer design in the view of upcoming Internet of Things (IoT) standards and their ubiquitous presence. In this work, we present a novel class of antenna which tries to address this gap and illustrates the design process toward a possible pathway to IoT

M.R. Tripathy (✉) · P. Ranjan · A.K. Singh
ASET, Amity University Uttar Pradesh, Noida, Uttar Pradesh, India
e-mail: mrtripathy@amity.edu

A.K. Singh
Tata Teleservices, Main Mathura Road, New Delhi, India

© Springer International Publishing Switzerland 2016
P.J. Soh et al. (eds.), *Theory and Applications of Applied Electromagnetics*,
Lecture Notes in Electrical Engineering 379, DOI 10.1007/978-3-319-30117-4_11

119

and related class of operations in 2.4 and 5.8 GHz. We believe that a robust and scalable physical layer design will lead to smooth communication process and hence stable performance of swarms participating in the dynamical boundary process as mobility, geographic coordinates, and sensing data overloads increase along with stringent security requirements.

A low-cost planar antenna for a robot has been reported in [1], while report of an interesting antenna design for an untethered microrobot is there in [2]. Multiple antenna-based robotic localization has been designed in [3], and a full-on-communication mechatronic system has been described in [4]. A conformal, structurally integrated antenna for flapping-wing robots has been designed in [5], and hybrid antenna has been reported in [6, 7]. A reconfigurable microstrip patch antenna has been reported in [8], and [9] discusses a localization system based on high-frequency antenna. Millimeter-wave harmonic sensors have been demon-strated in [10], and configurable robotic millimeter-wave antenna facility has been proposed in [11]. Many more interesting versions of antenna-based communication schemes for individual and swarming commutation can be found in [12, 13].

This work is motivated by the lack of multiband antennas for communication systems geared toward robots. To precisely bridge this gap, this work describes a multiband antenna for static and mobile robotic applications. A novel feature of this design is a hybridization of meander line resonator and split ring resonator struc-tures which provide multiband characteristics including ISM, IoT, and white space frequency bands. Simulation results in radiation patterns at different frequencies are presented.

This work is organized as follows: Sect. 2 is the process of basic antenna design. Section 3 contains the details of results and discussion of the proposed antenna, and Sect. 4 presents the conclusive remarks.

## 2 Antenna Design for Swarming Robots

### 2.1 Scenario Under Development

Multiple defense and academic agencies are working on an effort where technology for cooperative tracking of moving boundaries is being developed using evolvable curves and particular agile hardware is being developed to achieve this mission. There are algorithms proposed where an idea based on hybrid-level set has been used to achieve dynamic perimeter surveillance within a region by constructing an evolving function based on the perceived density of a phenomenon. The utility of this tech-nology is in providing surveillance, security cover, monitoring, tracking, etc., due to its versatile nature. This effective nature of boundary monitoring technology lends itself naturally toward "through-the-door" scene where two robots can be deployed to provide security cover as a part of overall team. While this scenario looks exciting at design table, non-trivial efforts are required to achieve the agile communication capability in the event of changing scenarios and dynamic overlay network.

**Fig. 1** Geometry of the
proposed antenna

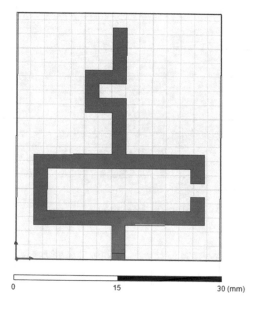

0                              15                              30 (mm)

## 2.2  Proposed Antenna Design

The antenna is designed on FR4 substrate with a thickness of 1.6 mm, $\varepsilon_r = 4.4$, and loss tangent 0.02. The top view of the antenna is shown in Fig. 1. The dimension of the antenna is $35 \times 30 \times 1.6$ mm$^3$. Microstrip-line feeding is used in this antenna. It has a structure which is a combination of split ring resonator and meander line. This design is simple, compact, and miniaturized. It seems promising to integrate with the rest of the dynamic boundary tracking robotic circuit. The antenna is simulated by using HFSS 14 (High-Frequency Simulation Software) by Ansoft.

## 3  Results and Discussion

The return loss versus frequency plot for the proposed antenna is shown in Fig. 2. Six interesting bands are obtained in this design. The band at 2.015 is quite broad with the impedance bandwidth of 1.52 GHz. It has a return loss of −23.13 dB. This band is of great interest for IoT and high-speed communication. The band at 0.66 GHz is very important, and it covers white space (470–698 MHz) and communication bands such as 860 MHz with the impedance bandwidth of 0.55 GHz and return loss of −20.90 dB. Third band at 4.25 GHz has the small impedance bandwidth of 0.123 GHz and return loss of −11.37 dB. Fourth band at 5.13 GHz has the impedance bandwidth of 0.368 GHz and return loss of −15.25 dB. Fifth band at 6 GHz has the impedance bandwidth of 0.369 GHz with the return loss

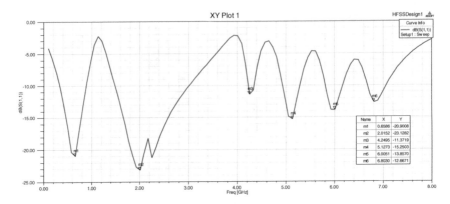

**Fig. 2** S11 parameter of the antenna

**Table 1** Antenna parameters at different frequency bands

| Sl. no. | Frequency (GHz) | Return loss (dB) | Bandwidth (GHz) | VSWR |
|---------|-----------------|------------------|-----------------|------|
| 1 | 0.66 | −20.90 | 0.55 | 1.22 |
| 2 | 2.015 | −23.13 | 1.52 | 1.18 |
| 3 | 4.25 | −11.37 | 0.12 | 1.78 |
| 4 | 5.13 | −15.25 | 0.37 | 1.5 |
| 5 | 6.00 | −13.86 | 0.37 | 1.5 |
| 6 | 6.80 | −12.67 | 0.38 | 1.65 |

of −13.86 dB. Finally, there is a sixth band at 6.8 GHz with the impedance bandwidth of 0.384 GHz and return loss of −12.67 dB. The details of these frequency bands are mentioned in Table 1.

Table 1 describes the return loss, impedance bandwidth, and VSWR of the proposed multiband antenna. Six interesting bands are available with this antenna. The maximum impedance bandwidth of 1.52 GHz is seen at 2.015 GHz frequency which is directly relevant to Internet of Things (IoT) kind of robotic and swarming applications.

Figure 3 shows the VSWR versus frequency plot for RFID tag antenna. The bands at 0.66, 2.0152, 4.25, 5.13, 6.00, and 6.8 GHz display VSWR 1.22, 1.18, 1.78, 1.5, 1.5, and 1.65, respectively. This is also mentioned in Table 1.

The simulated $E$ plane radiation pattern of the proposed multiband antenna at 738 MHz is shown in Fig. 4a. The pattern for $\Phi = 0$ is seen as omnidirectional, but the pattern at $\Phi = 90$ is fully bidirectional. Figure 4b illustrates H plane radiation pattern at 738 MHz. The radiation pattern at $\theta = 0$ is omnidirectional, and the gain is better than the E plane pattern. The radiation pattern at $\theta = 90$ is bidirectional with improved gain. In both the cases, radiation patterns show similar characteristics.

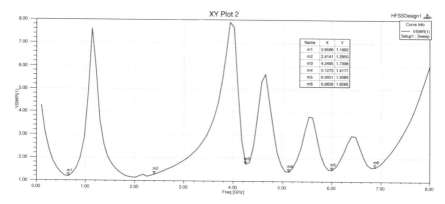

**Fig. 3** VSWR versus frequency plot of RFID tag antenna

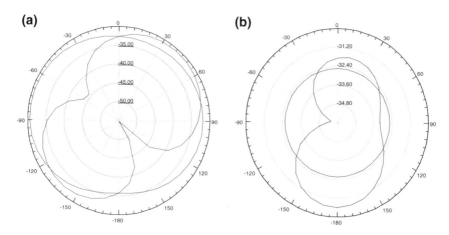

**Fig. 4** **a** $E$ field radiation pattern at 738 MHz and **b** $H$ field radiation pattern at 738 MHz

In Fig. 5a, b, the simulated $E$ and $H$ plane radiation patterns of the proposed multiband antenna at 2.414 GHz are shown, respectively. The patterns for $\Phi = 0$ and 90 are seen to be nearly omnidirectional. Both the patterns show similar features. The gain in both the cases is better than Fig. 4a, b. In Fig. 5b, it is shown that the radiation pattern for $\theta = 0$ is omnidirectional and for $\theta = 90$, it is directional. But the relative gain is better in comparison with corresponding pattern at 738 MHz.

The simulated $E$ and $H$ plane radiation patterns of the proposed multiband antenna at 5.845 GHz are shown in Fig. 6a, b, respectively. The patterns for $\Phi = 0$ and 90 are seen to be directional. But the gains in both the cases are better than the patterns shown in Figs. 4a, b and 5a, b, respectively. In Fig. 6b, radiation pattern

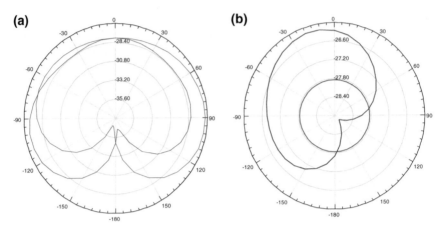

**Fig. 5** **a** E field radiation pattern at 2.414 GHz and **b** H field radiation pattern at 2.414 GHz

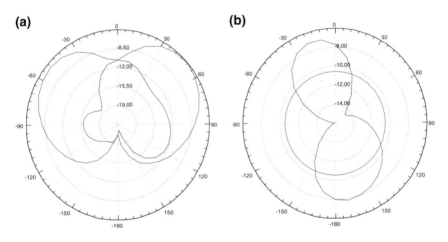

**Fig. 6** **a** E field radiation pattern at 5.845 GHz and **b** H field radiation pattern at 5.845 GHz

for $\theta = 0$ is omnidirectional and for $\theta = 90$, it is bidirectional. But the relative gain is better in comparison with the rest of patterns shown in this paper.

Figure 7 shows the surface current density distribution on radiating patch at 2.4 GHz. It is seen that current density is distributed throughout the radiating patch. The coupling between feedline and rest of the patch is very strong. However, the current distribution is stronger at the lower part of the split ring resonator region of the patch. The gain and other parameters can be improved by adjusting design parameters of antenna and ground plane of the proposed, compact, simple, and miniaturized multiband antenna.

**Fig. 7** Current density
distribution at 2.4 GHz

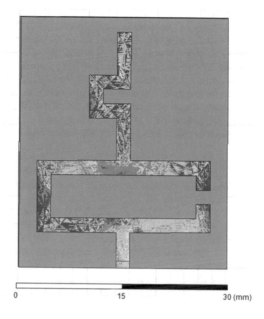

0                    15                    30 (mm)

## 4 Conclusions

This paper has addressed the issues of designing simple, compact, miniaturized, and multiband antenna for swarming robotic applications and their integration for multiband operation. It bridges the much needed gap for multiband antennas to support communication for robotic platforms. The antenna is designed on FR4 substrate with a thickness of 1.6 mm, $\varepsilon_r = 4.4$, and loss tangent 0.02. Future work includes real-life demonstration with this antenna integrated with the state-of-the-art robotic/vehicle platforms and next-generation antenna design from lessons learnt from these experiments.

## References

1. Perotoni MB, Garibello BE, Barbin SE (2006) An IEEE 802.11 low cost planar antenna for a mobile robot. In: IEEE antennas and propagation society international symposium 2006, pp 969–972, 9–14 July 2006
2. Martel S, Andre W (2009) Embedding a wireless transmitter within the space and power constraints of an electronic untethered microrobot. In: Joint IEEE North-East workshop on circuits and systems and TAISA conference, 2009. NEWCAS-TAISA '09, pp 1–4, 28 June 2009–1 July 2009
3. Sun Y, Xiao J, Cabrera-Mora F (2009) Robot localization and energy-efficient wireless communications by multiple antennas. In: IEEE/RSJ international conference on intelligent robots and systems, 2009. IROS 2009, pp 377–381, 10–15 Oct 2009

4. Bezzo N, Griffin B, Cruz P, Donahue J, Fierro R, Wood J (2014) A cooperative heterogeneous mobile wireless mechatronic system. IEEE/ASME Trans Mechatron 19(1):20–31
5. Oh J, Lee K, Forrest SR, Sarabandi K (2013) Conformal, structurally integrated antenna with a thin-film solar cell array for flapping-wing robots. In: 2013 IEEE antennas and propagation society international symposium (APSURSI), pp 1332–1333, 7–13 July 2013
6. Li K, Akbas MI, Turgut D, Kanhere SS, Jha S (2014) Reliable positioning with hybrid antenna model for aerial wireless sensor and actor networks. In: 2014 IEEE wireless communications and networking conference (WCNC), pp 2904–2909, 6–9 April 2014
7. Ju Y, Jin Y, Lee J (2014) Design and implementation of a 24 GHz FMCW radar system for automotive applications. In: 2014 international radar conference (Radar), pp 1–4, 13–17 Oct 2014
8. Kishan Kumar K, Prasanth ES (2015) A novel design approach and simulation of frequency reconfigurable micro strip patch antenna for Wi-Fi, WLAN and GPS applications. In: 2015 international conference on robotics, automation, control and embedded systems (RACE), pp 1–4, 18–20 Feb 2015
9. Dobrev Y, Vossiek M, Shmakov D (2015) A bilateral 24 GHz wireless positioning system for 3D real-time localization of people and mobile robots. In: 2015 IEEE MTT-S international conference on microwaves for intelligent mobility (ICMIM), pp 1–4, 27–29 April 2015
10. Tahir N, Brooker G (2015) Toward the development of millimeter wave harmonic sensors for tracking small insects. IEEE Sens J 15(10):5669–5676
11. Guerrieri JR, Gordon J, Novotny D, Francis M, Wittmann R, Butler M (2015) Configurable robotic millimeter-wave antenna facility. In: 2015 9th European conference on antennas and propagation (EuCAP), pp 1–2, 13–17 April 2015
12. Pukach A, Dupak B (2015) Development of reconfigurable antennas design model for mobile robotic design system. In: 2015 13th international conference the experience of designing and application of CAD systems in microelectronics (CADSM), pp 207–210, 24–27 Feb 2015
13. Say S, Aomi N, Ando T, Shimamoto S (2015) Circularly multi-directional antenna arrays with spatial reuse based MAC for aerial sensor networks. In: 2015 IEEE international conference on communication workshop (ICCW), pp 2225–2230, 8–12 June 2015

Printed in the United States
By Bookmasters